THE HURRICANE HUNTERS

Books by Ivan Ray Tannehill

HURRICANES; THEIR NATURE AND HISTORY

PREPARATION AND USE OF WEATHER MAPS AT SEA

WEATHER AROUND THE WORLD

DROUGHT; ITS CAUSES AND EFFECTS

ALL ABOUT THE WEATHER

THE HURRICANE HUNTERS

THE *Hurricane Hunters*

BY Ivan Ray Tannehill

ILLUSTRATED WITH PHOTOGRAPHS

DODD, MEAD & COMPANY
NEW YORK 1961

EIGHTH PRINTING

Library of Congress Catalog Card Number: 55-9480

Printed in the United States of America
by The Cornwall Press, Inc., Cornwall, N. Y.

To my daughter and son-in-law,
Doris and Bill

Acknowledgment

At appropriate places in the book the narrative serves as an acknowledgment by giving the names of a large number of men who furnished information in personal interviews, by correspondence, or in their reports which were included in the voluminous files searched in the last year.

In writing this book I had unstinted cooperation from the Air Weather Service and its Commander, Brigadier General Thomas Moorman, from the Aerological Branch of the Navy Department and its Head, Captain C. J. S. McKillip, and from the Chief of the Weather Bureau, Dr. F. W. Reichelderfer, and his associates in the field and the central office. In particular, Major William C. Anderson and associates in the Office of Information Services of the Air Weather Service and Captain Robert O. Minter of the Fleet Weather Central at Miami and his associates there in Airborne Early Warning Squadron Four at Jacksonville were extremely helpful. Of the associates of these men I wish to mention especially the assistance of Lieutenant Commander R. W. Westover and Air Force Captain Ed Vrable, both of whom are seasoned hurricane hunters.

Others not mentioned in the book who contributed to the

warning service and indirectly to the material used here were Isaac M. Cline and Charles L. Mitchell of the Weather Bureau. Their writings supply much of the background for any work on tropical storms.

The Air Force, Navy and Weather Bureau kindly supplied official photographs used here, except the wave breaking on the sea wall by the Miami *Daily News* and the drawings of sailing ships in hurricanes which are credited to Colonel William Reid who published them in 1850 in his book on the "Law of Storms."

The Author

CONTENTS

ILLUSTRATIONS

(Photographic supplement follows page 50)

THE HURRICANE HUNTERS

1. MONSTERS OF THE WORLD OF STORMS

The hollow winds begin to blow,
The clouds look black, the glass is low.
—E. Darwin

A stiff breeze, now and then with a hard gust, swept rain across the Navy airfield. The place was gloomy and deserted, except for one Privateer standing behind the air station, all other planes having been evacuated the night before. A tall young airman came out of a building down at the other side of the field. He looked nervously at the blackening morning sky as another squall came by, hurried over to the plane and stood between it and the protecting station. In a few minutes, eight men followed him. They climbed aboard the craft. The tall airman was last, taking a final look at the sky over his shoulder as he crawled in. The roots of his hair felt electrified, his spine tingled and his knees turned to rubber. In a few moments the plane took off into the darkening sky.

In those anxious moments as he had glanced upward at the wind-torn clouds with driving rain in his face, many thoughts passed through his mind. In training for this job he had read about aircraft carriers having their flight decks torn up by typhoons, about battered destroyers sunk by hurricanes, big freight ships tossed out on dry land, upper stories of brick buildings sliced off, timbers driven endways through the tough trunks of palm trees. The idea of sending a plane into one of these monsters seemed fantastic. He could imagine the wings being torn off and see vividly in his mind the broken craft rocketing downward into the foam of gale-swept waters far below. He leaned over on the radio table and muttered a prayer, hoping that God could hear him above the tumult of winds, seas and engines. To most of the men this was "old stuff." Flying into hurricanes had been going on for two years. To him it was a strange adventure.

He was the radio man and this was to be his first flight into a hurricane. And it would be no practice ride. This was a bad storm, getting too close to the coast to suit him. He had been told that after nightfall its center would strike inland and there would be widespread damage and some loss of life. He tried to remember other things they had told him in the briefing session and some of the instructions he had been reading for three days now. Well, such is life, he thought. His father had been the master of an oil tanker for the last fifteen years. He had told his growing son a lot about these big storms of the Caribbean. What would his father say now when he learned that his son was one of the men assigned to the job of flying into them? His thoughts were interrupted by violent agitation of the plane and the roar of the wind. The navigator said something about the turbulence.

He remembered asking one of the men what it would be like in the hurricane, and the fellow laughed and said, "Like going over Niagara Falls in a telephone booth." He recalled the burly fellow who pointed to the map and told them where the center of the hurricane was located and how to get to it. In answer to his last question, one of the men had told him that all he had to do was hold on for dear life with both hands until the weather officer handed him a message for the forecast office and then he should send it as quickly as possible, without being thrown on his ear. Now the plane was bumping along in the overcast and the rain had become torrential. The wind was on the port quarter and water was coming through the nose and flooding the crawlway. It was pouring on him from above somewhere. Rivers were running down his back.

He asked the weather officer what he thought about it, and he replied, "Oh, this is the usual thing. Sometimes it gets a good deal worse." Well, he thought it was getting a lot worse. Maybe the pilot and co-pilot could see but he could see nothing outside the plane. He hit his head on something, a hard crack, and he started to feel sick. Finally, he put his head down on the edge of the table and began to lose his breakfast.

Up and down the coast the Air Force bases were deserted. All planes but one had been flown inland and the last one, a B-17, was poised on Morrison Field for the final hop into the big winds, to return before nightfall.

In Miami, one of the senior men in the Weather Bureau office was called to the telephone. Somebody insisted on talking to him and nobody else. It was long distance. A woman said in a frightened voice that her son had gone out to look after a neighbor's boat and she wanted to know whether she should try to go out to find him and bring him in. He was only twelve years old. "Yes, by all means," was the answer. The forecaster didn't know how she was going to reach the boy or how far she had to go, but he recalled that other men and boys had lost their lives doing the same thing. They were having hundreds of calls and they were unable to go into details. He paused just a moment, his mind running regretfully over this poor woman and her problem. Then he started a radio broadcast.

Down the street, a merchant was pacing up and down on the sidewalk, bossing three men who were nailing frames over his plate glass windows. He went into the store to his telephone and, after dialing for about ten minutes, finally got the forecaster on the line. "What's the latest on the storm?" he asked in a strained voice. "Nothing new," came the tired voice of the forecaster. "A Navy plane went out half an hour ago. We'll have a report pretty soon now. But the hurricane's going to hit us, that's sure. Be a bad night."

Three miles south of the city, two fishermen stood looking at a pole on the pier. Two red flags with black centers were flapping in the wind. "Aw, nuts," growled the big man. "Guess I'll go home and nail up the windows again. This is the third time this year." The little man started off, pulling his raincoat up around his ears as a squall came over. "Well, we can't complain, I guess. The other times the flags went up we got storms, didn't we? Looks like this will be the worst of the lot." By that time the big fellow was running in a dog-trot and disappearing around a building. His father had been drowned in the big storm at Key West in 1919.

Even on the other side of the State the people were worried, and for good reason, for it might be over there tomorrow. The forecaster was wanted again on the telephone. A man said in an anxious tone that he had one thousand five hundred unfenced cattle near the shore and what should he do? Without hesitation, the forecaster said, "Get them away from the water and behind a fence. This storm will go south of you. There will be strong offshore gales and the cattle will walk with the wind and go right out into the water and drown if there is no fence."

Out in the Atlantic, a merchant ship was wallowing in

heavy seas, with one hundred miles an hour winds raking her decks. The third mate struggled through the wind and sea and into the radio room. He handed a wet weather message to the radio operator. A hundred miles away, in the Bahamas, an old Negro was reading his weather instruments and looking at the sky. He was pushed around by furious winds but they had died down a little since early morning. The roof was off his house. Trees were uprooted all around him. He went into a small, low-slung radio hut and attempted to send a weather message to Nassau. He was badly crowded in the hut. His wife, daughter and two grandchildren were huddled in the corners. His son-in-law had been killed in the night by a big tree that fell on the porch. His daughter and her two children were sobbing. He raised the Nassau radio station and sent a message for the forecast office in Miami.

All up and down the Florida coast, many thousands had heard the radio warnings or had seen the flags flying and wanted to know more. The highways here and there were filling with people, leaving threatened places on the coast. By night the roads would be jammed. Out on the Privateer, the tall young radioman, sopping wet, raised himself in his chair, and took a soggy message from the weather officer. After the plane settled a little, he put on his head phones and listened to the loud, almost deafening static. He still felt a bit sick. But he began to pound out the weather message, with the hope that somebody would get it and pass it on to the forecaster.

In these and other ways, it has come about that a pair of red flags with black centers strikes fear into the hearts of seafaring men and terrifies people in towns and cities in the line of advance of the big winds. The warning brings to their minds raging seas and screaming gales, relatives and friends lost in other great storms that have roared out of the tropics, ships going down and buildings being torn apart.

Ahead of the storm, the sea becomes angry. Huge rollers break on the beaches with a booming sound. In the distance, a long, low, angry cloud appears on the horizon. If the cloud grows and puts out scud and squalls, spitting rain, the warning flags flutter in the gusts and the big winds will strike the coast with terrible destruction. If the distant cloud is seen to move along the horizon, the tumult of wind and sea on the beaches will subside. The local indications in the sky and the water tell a vital story to the initiated but the warning they give does not come soon enough. It is necessary to know what is going to happen while the hurricane is well out at sea. This depends on the hurricane hunters, and so the messages they send ashore while fighting their way by air into the vortices of these terrible whirlwinds are awaited anxiously by countless people.

Tracking and predicting hurricanes is an exciting job, often a dangerous one. But it is not a one-man job; it requires the co-operation of many people. A tropical storm of hurricane force covers such a vast area that all of it cannot be seen by one person. Its products—gales with clouds and rain—and its effects—destruction of life and property and big waves on the sea—are visible to people in different parts of the disturbance. But before we know much about it, the little that is seen by each of many people on islands and ships at sea must be put together, like clues in a murder case. The weather observers who get the clues and the experts who put them together are the hurricane hunters.

For at least five hundred years it has been known that these terrible disturbances are born in the heated parts of the oceans. Down near the equator, where hot, moist winds are the rule, something causes vast storms to form and grow in violence, bringing turmoil to the ordinary daily round of gentle breezes and showers. They have come to bear the general name of tropical storms, though known locally as hurricanes, typhoons, or cyclones.

Most of them occur in the late summer or early fall. At that season, on the islands in the tropics where the natives in other centuries took life easy, depending on nature's lavish gifts of fruit and other foods, the tropical storm came as an occasional catastrophe. Trees went down in howling gales, rain came in torrents, flooding the hilly sections, big waves deluged the coasts, and frail native houses were swept away in an uproar of the elements. The survivors thought they had done something to displease one of the mythical beings who ruled the winds and the waters. In the Caribbean region, it was supposed to be the god of the big winds, Hunrakan, from which the name hurricane originated. His evil face seemed to leer from the darkening clouds as the elements raged.

In time, Europeans settled in the islands and on the southeastern coasts of America. They dreaded the approach of late summer, when copper-colored clouds of a tropical storm might push slowly upward from the southeastern horizon. What they learned about them came mostly from the natives, who had long memories for such frightening things and reckoned the time of other events from the years of great hurricanes. Strangely enough, although during the more than four hundred years that have passed since then, man has finally mastered thermo-nuclear reactions capable of permanent destruction of whole islands, he still probes for the secret of storm forces of far greater power.

It is hard to say who was the first hunter of storms. Columbus and his sailors were constantly on the lookout and actually saw several West Indian hurricanes. Luckily, they didn't run into one on their first voyage, or the story of the discovery of America would be quite different, for the ships sailed by Columbus were not able to stand up against these big winds of the tropics. They would have been sunk in deep water or cast ashore as worthless wrecks.

If Columbus had been lost in one of these monstrous storms—and he didn't miss it by very much—it might have been many years before another navigator with a stout heart could have induced men to risk their lives in the uncharted winds of the far places in the Atlantic Ocean. Out there toward the end of the world, where increasing gales dragged ships relentlessly in the direction of the setting sun, sailors who ventured too far would drop off the edge of a flat earth and plunge screaming into eternity—so they thought. Only in Columbus' mind was the earth a sphere.

By the time Columbus had made his third voyage to the West Indies, he had learned a good deal about hurricanes and how to keep out of them. He got this information by his own wits and from talking with the natives in the islands bordering the Caribbean. They told him of storms much more powerful than any that were brewed in European waters. After listening to their tales, he was afraid of them. In 1494 he hid his fleet behind an island while a hurricane roared by. The next year, an unexpected one sank three of his vessels and the others took such a beating that he declared, "Nothing but the service of God and the extension of the monarchy would induce me to expose myself to such dangers."

In 1499, a Spaniard named Francisco Bobadilla was appointed governor and judge of the Colony on Hispaniola (Santo Domingo). He sent false charges back to Spain, accusing Columbus of being unjust and often brutal in his treatment of the natives. Columbus was ordered back to Spain in chains. Here he remained in disgrace until December, 1500. By that time the true nature of Bobadilla's treachery had become known.

By the spring of 1502, Columbus had been vindicated and was on his way back to the West Indies with four ships and 150 men. During his earlier voyages he had become deeply respectful of these big winds of the New World. When he arrived at San Domingo on this last voyage, his observations made him suspect the approach of a hurricane. At the same time, a fleet carrying rich cargoes was instructed to take Bobadilla back to Spain. It was ready to depart. Columbus asked for permission to shelter his squadron in the river and he sent a message, urging the fleet to put off its departure until the storm had passed.

Bluntly, both of Columbus' requests were denied. He found a safe place in the lee of the island but the fleet carrying Bobadilla departed in the face of the hurricane and all but one vessel went to the bottom. Bobadilla went down with them, which seemed to be a fitting end for the scoundrel who had been guilty of hatching up false charges against Columbus.

After the time of Columbus, better ships were built and the fear of storms diminished. Seafaring men today are likely to get the idea that modern ships of war and trade are immune to hurricanes. They have a brush or two with minor storms or escape the worst of a larger one and cease to be afraid of the big winds of the West Indies. Now and then this attitude leads to disaster.

In September, 1944, the Weather Bureau spotted a violent storm in the Atlantic, northeast of Puerto Rico. It grew in fury and moved toward the Atlantic Coast of the United States. The forecasters called it the "Great Atlantic Hurricane." Being usually conservative, Weather Bureau forecasters seldom use the word "great" when warning of hurricanes and when they do, it is time for everybody to be on guard. In this case, the casualties at sea included one destroyer, two Coast Guard cutters, a light vessel and a mine sweeper. This should have been sufficient evidence of the power of the tropical storm to destroy modern warships, but just three months later a big typhoon caught the Navy off guard in the Pacific and proved the case beyond the slightest doubt.

Typhoons are big tropical storms, just like West Indian hurricanes. They form in the vast tropical waters of the Pacific, develop tremendous power, and head for the Philippines and China, sometimes going straight forward and sometimes turning toward Japan before they reach the coast. Like hurricanes, they are often preceded by beautiful weather, allaying the suspicions of the inexperienced until it is too late to escape from the indraft of the winds and the mountainous seas that precede their centers.

It was hard to keep track of typhoons in World War II. In large areas of the Pacific there are few islands to serve as observation posts for weathermen. Before the war, merchantmen on voyages through this region had reported by radio when they saw signs of typhoons. But many of the weather-reporting vessels had been sent to the bottom by enemy torpedoes and the remainder had been ordered to silence their radios. Thereafter, the only effective means of finding and tracking tropical storms was by aircraft, but reconnaissance by air had just begun in the Atlantic and was not organized in the Pacific until 1945.

Late in 1944, our Third Fleet, said to be the most powerful sea force ever assembled, had drawn back from the battle of Leyte to refuel. The Japanese Navy had received a fatal blow from the big fleet. Nothing more terrible was reserved for the Japanese except the atom bomb. Far out in the Pacific, a typhoon was brewing while valiant oil tankers waited five hundred miles east of Luzon for the refueling operation so vitally needed by our warships after days of ranging the seas against the Japs.

It was December 17 when the refueling began. By that

time, the winds and seas in the front of the typhoon were being felt in force. Battleships, cruisers, destroyers and a host of other vessels rode big waves as the wind increased. The typhoon drew nearer and the smaller ships were bounced around so violently that it became impossible to maintain hose connections to the oilers. Before nightfall, the refueling had stopped completely and the fleet was trying to run away from the typhoon.

It was almost a panic, if we can use the word to describe the desperate movements of a great battle fleet. Messages flew back and forth, changing the ships' courses as the wind changed. They ran toward the northwest, then toward the southwest, and finally due south, in a last effort to escape the central fury of the great typhoon. But all this did no good.

The lighter vessels, escort carriers, destroyers, and such, top-heavy with armament and equipment and with little oil for ballast, began the struggle for life. Each hour it seemed that the height of the storm had come, but it grew steadily worse. Writhing slopes of vast waves dipped into canyon-like depths. The crests were like mountains. The wind came in awful gusts, estimated at more than 150 miles an hour. The tops of the waves were torn off and hurled with the force of stone. Ships were buried under hundreds of tons of water and emerged again, shuddering and rolling wildly.

On the eighteenth of December, one after another of the ships of the Third Fleet lost control and wallowed in the typhoon. Time and again thousands of men faced death and escaped by something that seemed a miracle. There was no longer any visible separation between the sea and the atmosphere. Only by the force with which the elements struck could the men aboard distinguish between wind-driven spume and hurtling water. Steering control was lost; electric power and lights failed; lifeboats were torn loose; stacks were ripped off; planes were hurled overboard; three destroyers rolled too far over and went to the bottom of the Pacific.

Altogether, nearly 150 planes were destroyed on deck or blown into the sea and lost. Cruisers and carriers suffered badly. Battleships lost planes and gear. The surviving destroyers had been battered into helplessness. Almost eight hundred men were dead or missing. As the typhoon subsided, the crippled Third Fleet canceled its plans to strike against the enemy on Luzon and retreated to the nearest atoll harbor to survey its losses. More men had died and more damage had been done than in many engagements with the Japanese Navy.

A Navy Court of Inquiry was summoned. It was said that this typhoon of 1944 was the granddaddy of all tropical storms. But a study of the records shows that it was just a full-grown typhoon. There have been thousands of hurricanes and typhoons like this one. Down through the centuries, these terrible storms have swept in broad arcs across tropical waters, reaching out with great wind tentacles to grasp thousands of ships and send them to the bottom. Pounding across populous coasts, with mountainous seas flooding the land, they have drowned hundreds of thousands of people, certainly more than a million in the last three centuries, and untold thousands before that.

After the typhoon disaster, the Commander-in-Chief of the Pacific Fleet declared that his officers would have to learn forthwith about the law of storms. Really there was nothing new in that idea. It had been voiced by navigators of all maritime countries of the world from the earliest times. The so-called "law of storms" is merely the total existing knowledge about storms at sea—how to recognize the signs of their coming and how to avoid their destructive forces—

and it has taken four and a half centuries to develop our present understanding of hurricanes.

This experience of the Third Fleet made it plain that a sailing vessel had very little chance of survival in the central regions of a fully developed tropical storm. The only hope was that the master would see the signs of its coming and manage to keep out of it. Once he became involved, the force of the wind was likely to be so great that his vessel soon would be reduced to an unmanageable hulk. The gales seemed to have unlimited power. Even today, we don't know accurately the speed of the strongest winds. It seems likely that the highest velocities are between two hundred and two hundred and fifty miles per hour. Wind-measuring instruments are disabled or carried away and the towers or buildings which support them are blown down.

Long after the time of Columbus, it was generally believed that a storm was a large mass of air moving straight ahead at high velocities. A ship might be caught in these terrible winds and be carried along with them, to be dashed on shore or torn apart and sent to the bottom. Every mariner wanted to know how to avoid these dangers but, strangely enough, few wanted to avoid them altogether. If a sailing vessel circled around a storm, it took longer to get to the port of destination and how could the master explain the time lost to his bosses when he got home, if he had no record of a storm in the log book to account for the delay?

From this point of view, some of the things that happened seemed very strange. Two or three hundred years ago, it was not uncommon for a sailing ship to be caught in a hurricane and scud for hours or days under bare poles in high winds and seas, and finally come to rest near the place where it first encountered the storm. A sailor on board would imagine he had traveled hundreds of miles and yet he might survive the wreck of his ship and find himself tossed ashore near the place where he started!

Up until about 1700 A.D., nobody could offer a reasonable explanation of these curious happenings and most people believed they never would be accounted for. For example, it was often claimed that "the storm came back." After blowing in one direction with awful force until great damage had been done, it would suddenly turn around and blow in the opposite direction, perhaps harder than before, wrecking everything that had not been destroyed in the first blow. To add to the mystery, many ships were never heard from again. They became involved in hurricanes and disappeared, leaving no trace of any kind.

Men might try to explain what had happened to the ships which were tossed on shore near the places where they had started from, but there was a general feeling that these cases were the exceptions to the law of storms and that the true understanding of these fearful winds would come only with the discovery of what happened to the great numbers of ships and men that were never seen again. And yet it is amazing to find how near some of these men came to the right answer. There were seafaring men in the seventeenth century who knew or suspected the truth but none of them had both the knowledge and the ability to put it in writing in a convincing manner. They were the buccaneers whose operations were centered in the Caribbean Sea, mostly from about 1630 to 1690. They were English, Dutch, Portuguese and French, all at one time or another opposed to Spanish control in the Carribbean. On various occasions they seized one or another of the smaller islands and used it as a base from which to prey on Spanish shipping and settlements.

During these years, the islands were devastated by at least thirty hurricanes of sufficient power to earn a place in history. Doubtless, there were many more not recorded. A great number of vessels went down in the seas and harbors around St. Kitts, Martinique and Jamaica, where the buccaneers sought haven from the Spaniards.

One of the most intelligent but least successful as a buccaneer was William Dampier. He was born in England in 1652, became an orphan at an early age and was put in the hands of the master of a ship in which he made a voyage to Newfoundland. Afterward, he sailed to the East Indies and then fought in the Dutch War in 1673. The next year he went to Jamaica and became a buccaneer. Soon he was familiar with the harbors, bays, inlets and other features of the Carribbean coasts and islands. At times, he and other buccaneers ranged as far as the South American coast, plundering, sacking and burning as they went. Eventually, they raided the Mexican and Californian coasts and crossed the Pacific to Guam, and then to the East Indies.

At intervals, Dampier wrote the accounts of his voyages which ultimately took him over most of the world. But he died poor, just three years before he was due to share in nearly a million dollars' worth of prize money.

Being a genius at the observation of natural phenomena and having the ability to put this in writing, Dampier distinguished himself from the other buccaneers by earning a place in history as a writer of scientific facts in a clear and easy style. In his writings, we find our earliest good firsthand descriptions of tropical storms that are really good. Among other things, he said of a typhoon in the China Sea that "typhoons are a sort of violent whirlwinds." He said they were preceded by fine, clear and serene weather, with light winds.

"Before these whirlwinds come on," wrote Dampier, "there appears a heavy cloud to the northeast which is very black near the horizon, but toward the upper part is a dull reddish color." To him, this cloud was frightful and alarming. He went on to say that it was sometimes seen twelve hours before the whirlwind struck. The tempest came with great violence but after a while the winds ceased all at once and a calm succeeded. This lasted an hour, more or less, then the gales were turned around, blowing with great fury from the southwest.

These stories by Dampier and others might have cleared up some of the mysteries of these furious storms, especially those that "turned around and came back." They might have explained the fact that sailors were carried long distances and then cast ashore near the places from which they started —for they were huge whirlwinds, as Dampier suspected—but nobody seemed to be able to put "two and two together" and prove it. For one thing, no one knew then that weather moves from place to place. Everybody seemed to have a vague belief that the weather developed right at home and blew itself out without going anywhere. With these ideas in vogue, the eighteenth century came to an end and there was no useful law of storms. But we can put William Dampier down as one of the first "hurricane hunters."

As cities and towns on southern coasts and islands grew in population, storm catastrophes became more numerous. Now and then, a hurricane seemed to appear from nowhere and caused terrible destruction on land. New Orleans was devastated in 1722 and again in 1723. Charleston and other coastal cities were hit repeatedly. Coringa, on the Bay of Bengal, was practically wiped out by a furious storm in December, 1789, and there was another disaster at the same place in 1839.

Tropical storms that form in the Bay of Bengal and strike the populous coasts of India are known as cyclones. They are the same kind of storms as West Indian hurricanes and the typhoons of the Pacific. The worst feature is the overwhelming flood of seawater that comes in big waves into the harbors as the center of the storm arrives. If there is insufficient warning, thousands of the inhabitants are drowned.

Coringa is a coastal city of India which had a population of about 20,000 in 1789. In December, there was a strong wind, "seeming like a cyclone." The tide rose to an unusual height and the wind increased to great fury from the northwest. The unfortunate inhabitants saw three huge waves coming in from the sea while the wind was blowing with its greatest violence. The first wave brought several feet of water into the city. All the able-bodied ran for higher ground or climbed to the rooftops to keep from drowning. The second wave flooded all the low parts of the city and the third overwhelmed everything and carried the buildings away. All the inhabitants, except about twenty, disappeared.

In cases of this kind, a warning less than an hour in advance would have saved the lives of thousands, but disasters like this were repeated here and in other parts of the world dozens of times before the hunters, trackers and forecasters of hurricanes learned to cheat these terrible storms of their toll of death and injury. Progress was slow in the nineteenth century, which saw some of the world's worst storm disasters. In 1881, three hundred thousand people died in one typhoon on the coast of China.

We now come to the stories of the men who tried to do something about it—the storm hunters. At first, early in the nineteenth century, the hunters were men engaged in some other work for a living. They put in their spare time gathering information, getting reports from sailors who had survived these terrible storms at sea and from landsmen who had seen them come roaring across harbors and beaches, to lay waste to the countryside. We go with some of them through these awful experiences. Then, after the middle of the century, first under Emperor Napoleon III of France and later under President Grant in America and Queen Victoria in England, storm hunting became a government job and spread slowly around the world.

Here we see a bitter uphill battle. The hurricane proved to be an enormous whirlwind, hidden behind dense curtains of low-flying clouds, tremendous rains, and the thick spray of mountainous seas torn by earth-shaking forces of the monster. Its mysteries were challenging. Out of this work a warning system grew, and slowly the losses of life were reduced from thousands to hundreds, and then to dozens. We go with the storm hunters into Congress and the White House, to argue about it. Then we come to World War II and the desperate need for information while submarines attack shipping and hurricanes threaten airfields and naval bases.

And here we find stories of big four-engined bombers flying into the centers of these furious storms. In these stories we go along. We see what the weather crews saw and learn what they learned. And we see how the hurricane warning service works today—far better than a few years ago—but with a part of the great mystery still unsolved. So we go with the hunters in shaking, plunging planes, from the surface of the sea to the tops of the biggest hurricanes, looking for the final answers to this great puzzle of the centuries.

2. THE SADDLER'S APPRENTICE

>>

"All violent gales or hurricanes are great whirlwinds."—Redfield

In recent years, when men were first assigned to the alarming duty of flying into hurricanes and they began to study the old records, one question bothered them very much. Why did it take so long to prove without doubt that these big tropical storms are whirlwinds? The main reason, of course, is the huge size of the wind circulation. The winds spiral in such a broad arc around the storm center that there is no noticeable change in the wind direction within a distance of many miles. It was like the curvature of the earth. Any circle around the full body of the earth is so enormous that it seems to be a straight line, and men were deceived for centuries into believing that the earth is flat.

The crews of fast modern aircraft can fly through the main part of a hurricane in two or three hours, at most, and they can immediately see changes of the wind as they go along. They have no reason to question it. In earlier times, there was no means of travel fast enough to get the facts in this way. Then, too, there was no means of sending messages fast enough to show what the wind was doing at the same instant in different parts of the storm. Also, the entire wind system was in motion and if the various reports were not sent at the same time, the results, when they were charted, failed to make sense. This fact alone was the cause of much confusion, even as late as the first part of the nineteenth century.

A definite answer to the whirlwind question came suddenly and unexpectedly in a most peculiar manner.

In the autumn of 1821, a young saddler was walking through the woods of central Connecticut with his inquiring mind on scientific matters of the day when he discovered a strange fact that led to the first "law of storms" and eventually made him the most illustrious of the hurricane hunters. His name was William Redfield. His ideas were first published in 1831 and, together with the work of a few men who followed on his trail, were the mainstay of sailors in stormy weather for nearly a hundred years.

Hurricanes were not only extremely dangerous to the sailing ships of that day but were becoming more destructive to the growing cities along the American coast. In the first quarter of the century, the population of the country doubled. In 1800, there were five million people. In spite of the War of 1812, which lasted for three years, and the temporary drop it caused in immigration, the population increased rapidly, mostly on and near the Atlantic Coast. The United States began to take a place in the forefront of the world's commerce. But now and then a great storm from the tropics swept the entire seaboard and took a grievous toll of ships and men and harbor facilities.

Up to that time, no one had learned enough about storms to give warnings in advance. There were no really useful rules to guide seamen around or out of a tropical storm. Weather prediction was not accepted as scientific work.

Storm disasters were called "acts of God" and the ways of the atmosphere were thought to be beyond human understanding.

Occasionally, a mariner with an inquiring mind like Dampier came to the conclusion that tropical storms are huge whirlwinds which move from place to place. But none of these inquirers came up with any real proof. After 1800, the destruction from hurricanes grew steadily worse. The summer of 1815 was remarkable for furious storms all along the Atlantic Coast. Newspapers were filled with the details of storm disasters and the destruction of life and property on shore and at sea. The crowning catastrophe was caused by a furious West Indian hurricane which struck New England on September 23 of that year. In the violence of its winds and the height of its tides, this storm was about equal to the New England hurricane of 1938. Although the country was far less populous in 1815, and the buildings, ships, and wharves subjected to its fury were much less numerous than in 1938, the destruction was so great and the loss of life so heavy that the newspapers did not have space enough to give all the details of the marine disasters in this instance.

At Providence, there was terrible destruction. The tide rose more than seven feet above the highest stage previously recorded. Five hundred buildings were destroyed; the loss of life was never fully determined, but it was excessive. The same sort of tragic story came from New Bedford and other towns on the coast. Many buildings and a tremendous number of trees were blown down in the interior.

The most treacherous feature of these big storms was their resemblance in the initial stages to the ordinary "northeasters" which came at about the same time of year—late August or September—and blew fitfully for a day or two. They brought rain and high tides along the coast and finally died out without much damage. Tropical storms, like the big one in 1815, begin much the same way in New England, but suddenly become violent. Then, as now, they blew gustily from the northeast in the beginning but went around the compass and ended with shattering on-shore gales which drove engulfing floods into the harbors. Everybody was caught off guard.

This storm and another which came six years later in the same region set men to thinking seriously about ways to avoid these disasters. The violent hurricane of 1821 crossed Long Island and New England, leaving a path of destruction which lay somewhat to the westward of the hurricane path of 1815. Again enormous numbers of trees were blown down, this time mostly in Connecticut. And here is where we come to the story of the saddler's apprentice.

In September, 1802, a sailor named Peleg Redfield, of Middletown, Connecticut, died, leaving a widow and six children in very poor circumstances. The eldest child, William, thirteen years of age, had attended common school and learned about reading, writing and arithmetic, but when his father died, he had to be taken out of school.

The next year William was apprenticed to a local saddle and harness maker. Boys as well as men worked long hours in those days, and William Redfield was no exception. After he had finished the day's work and had done the chores around the Redfield home, he had only a small part of his evening to himself. Even then, he had a lot of discouragement—no books and no light to read by. The family could not afford candles. Nevertheless, William was so interested in science that he studied by the light of the wood fire, reading intently anything on scientific subjects that he could get his hands on.

A year later, William's mother married a widower with nine children of his own, and in 1806 the couple moved to Ohio, taking his nine children and five of hers, but leaving

William behind to look out for himself. He continued his study of science, but with no indication that he would eventually find some of the answers so vitally needed in the fight against hurricanes. His father, being a sailor, had told him about storms at sea and the boy was unable to get this out of his mind.

Fortunately, there was a well-educated physician in the village of Middletown, William Tully, who had a good library and made it available to young Redfield. The first book the physician handed to William was a very difficult volume on physics. The boy brought it back so soon the doctor thought he had been unable to understand it, but he was pleasantly surprised, for the lad had read it very thoroughly and had come back for more technical works of the time. Soon William gained such an understanding of scientific matters that an intimate friendship with the physician developed. During this time, however, young Redfield felt an increasing urge to visit his mother. But she lived more than seven hundred miles from Middletown and he had very little money. So in 1810 he walked all the way to Ohio.

At that time, Ohio had a very small population; it was less than 50,000 at the beginning of the century. The territory intervening between Ohio and Connecticut was pretty wild, with settlements only here and there. William followed primitive roads and trails and at last reached the shores of Lake Erie, where Cleveland and other cities stand today. The next year he walked back to Connecticut.

Redfield was now past twenty-one. He had thought deeply of many things while he trudged those lonely trails. He had a vision of a great railway extending from Connecticut to the Mississippi River. Also, his mind kept running back over the stories of storms his father had told him. From his thoughts on this lonely journey he devised and later executed a plan for a line of barges which operated between New York and Albany.

But when he arrived in Middletown, he had no course for the time being except to go into business in his trade of saddler and harness maker. To supplement his poor income, he peddled merchandise in the region around Middletown, trudging through the woods and stopping in the villages here and there. The years went by and he kept on studying science in his spare moments.

And then, on the third of September, 1821, the center of that vicious hurricane which crossed the eastern part of Connecticut brought its dire evidence to the very door of the man who was still trying to master the sciences in his spare moments. As Redfield trudged the countryside with his wares, he passed among hundreds of big trees felled by the furious winds. Near Middletown, he found that the trees lay with their branches toward the northwest and he remembered that the gale there had begun from the southeast. Less than seventy miles away, he found the trees lying with their heads toward the southeast and here the winds evidently had begun from the northwest.

Making inquiries as he went along, Redfield learned the directions from which the winds had blown at various times during the storm. It became quite clear that the hurricane had been a huge whirlwind which had traveled across the country from south to north. He gathered a lot of evidence to prove it.

But Redfield was now past thirty years of age. Because he had not gone very far in school, he did not see how he could undertake to demonstrate these facts about hurricanes to men of scientific learning. He kept turning the idea over in his mind at intervals as the months and years went by. In the meantime, he had become interested in navigation on the Hudson River and had made a reputation as a marine engineer. By 1826, he was superintendent of a line of forty or fifty barges and canal boats. But whenever he read of a bad storm on the coast, he thought about the hurricane of 1821 and the trees thrown down in different directions by the opposing winds of a great whirling storm.

In 1831, Professor Denison Olmstead of Yale College was traveling by boat from New York to New Haven. A stranger approached him and began talking about some papers the professor had published in the *American Journal of Science.* The stranger said his name was William C. Redfield. (Actually he had no middle name but used the C for "Convenience," to keep from being confused with two other William Redfields in the area.) In the course of the conversation, Redfield talked reservedly about his ideas regarding West Indian hurricanes. The professor was amazed and urged him to publish his ideas in the *American Journal of Science.*

Redfield, who was now forty-two years old, began writing on the law of storms. He wrote well and his ideas were clear and convincingly expressed. A long series of articles followed his first one in the *American Journal of Science.* During these years he became a famous "hurricane hunter." He collected reports of West Indian hurricanes—as many as he could get from ships caught in storms and from other sources—and studied them at great length. He inspected the log books of vessels in port, interviewed many shipmasters, and corresponded with others. His urgent purpose was to devise a law of storms and a set of rules to promote the safety of human life and property afloat on the oceans and to afford some measure of protection for the inhabitants of cities and towns on the coasts subjected to destructive visits from these monsters of the tropics.

After the death of Redfield, in 1857, Professor Olmstead summarized his theory of storms as follows:

"That all violent gales or hurricanes are great whirlwinds,

in which the wind blows in circuits around an axis; that the winds do not move in horizontal circles but rather in spirals.

"That the direction of revolution is always uniform being from right to left, or against the sun, on the north side of the equator, and from left to right, or with the sun, on the south side of the equator.

"That the velocity of rotation increases from the margin toward the center of the storm. That the whole body of air is, at the same time, moving forward in a path, at a variable rate, but always with a velocity much less than its velocity of rotation.

"That in storms of a particular region, as the gales of the Atlantic or the typhoons of the China Sea, great uniformity exists with regard to the path pursued by these storms. Those of the Atlantic, for example, usually come from the equatorial regions east of the West India islands, moving at first toward the northwest as far as the latitude of 30°, and then gradually wheeling toward the northeast and following a path nearly parallel to the American Coast until they are lost in mid-ocean. That their dimensions are sometimes very great, as much as 1,000 miles in diameter, while their paths over the ocean can sometimes be traced for 3,000 miles."

These conclusions were in the main correct, but time has proved that there are many exceptions. At any rate, Redfield's papers became classics. He had demonstrated by collections of observations on shipboard that a tropical storm is an organized rotary wind system and not just a mass of air moving straightaway at high velocities.

It happened that in 1831, the same year in which Redfield's first paper appeared in the *American Journal of Science*, there was a terrible hurricane on the island of Barbados. Devastation was so great that the people on the island firmly believed the storm had been accompanied by an earthquake. More than 1,500 lives were lost. Property damage, considering values in that early day, was tremendous for a small island—estimated at more than seven million dollars.

Barbados had suffered so much that England sent Colonel (afterward Brigadier-General) William Reid of the Royal Engineers to superintend the reconstruction of the government buildings. He was appalled by what he saw.

Reid examined the ruins and made inquiries of many people about the nature of the hurricane of 1831. He came to the conclusion that there had not been an earthquake, but all the damage had been caused by the wind and sea. One of the residents told Reid that when daybreak came, amidst the roar of the storm and the noise of falling roofs and walls, he had looked out over the harbor and saw a heaving body of lumber, shingles, staves, barrels, wreckage of all description, and vessels capsized or thrown on their beam ends in shallow water. The whole face of the country was laid waste. No sign of vegetation was seen except here and there patches of a sickly green. Trees were stripped of their boughs and foliage. The very surface of the ground looked as if fire had run through the land.

Reid resolved to study hurricanes and see what he could do to reduce the consequent loss of life. He wanted to tell sailors how to keep out of these terrible storms and he thought it might be possible to design buildings capable of withstanding the winds. Soon afterward, he saw Redfield's articles in the *American Journal of Science.* He wrote to the author and they began a friendly correspondence which continued until the latter's death.

Neither Redfield nor Reid was actually the first to declare that the hurricane is a great whirlwind. Many others had suggested this before them, and in 1828 a German named H. W. Dove had confirmed it, but none of these had hunted up the data and talked and corresponded with hundreds of seamen to collect facts to prove their contentions. And none had presented the facts in a way that would serve as a law of storms for seamen.

Following the lead of General Reid, an Englishman named Henry Piddington, on duty at Calcutta in India, became a great hurricane hunter in the middle of the nineteenth century. He collected information from every source, talked to seamen of all ranks from admiral down, and added a great deal to the law of storms. Because of the movement of violent winds around and in toward the hurricane center, he gave it the name *cyclone,* which means "coil of a snake." This is the reason why tropical storms are now called cyclones in the Bay of Bengal.

Piddington, who became President of the Marine Courts of Inquiry at Calcutta, published numerous memoirs on the law of storms. Of all the accounts that he collected of experiences of seamen in tropical storms, the outstanding case, in his estimation, was that of the Brig *Charles Heddles,* in a hurricane near Mauritius, a small island in the Indian Ocean, east of Africa. Mauritius is south of the equator, where hurricane winds blow around the center in a clockwise direction, the opposite of the whirling motion of storms in the northern hemisphere.

The *Charles Heddles* was originally in the slave trade but at the time that she was caught in the hurricane was mostly being employed in the cattle trade between Mauritius and Madagascar. Only the fastest vessels were engaged in the cattle trade, and the *Charles Heddles* was an exceptionally good ship. Her master was a man named Finck, an able and highly respected seaman.

On Friday, February 21, 1845, the *Charles Heddles* left Mauritius and in the early morning of the twenty-second ran into heavy weather, with wind and sea gradually increasing. It became squally and the vessel was laboring greatly by midnight. On the twenty-third it was worse, with a frightful sea and the wind very high, accompanied by incessant rain. The seas swept over the decks and the crew was frequently at the pumps.

By this time Captain Finck had determined to keep the brig scudding before the wind and run his chance of what might happen. The steady change of the wind around the compass as the day wore on made it impossible for him to estimate his position, but he was sure he had plenty of sea room. The crew was unable to clue up the topsail without risk of severe damage, so round and round they went.

Wind force and weather were always about the same. There was a terrifying sea, the vessel constantly shipping water, which poured down the hatchways and cabin scuttle. The fore topsail blew away at 4 P.M. and they continued scudding under bare poles, the ship's course changing steadily around the compass. By the twenty-fifth of February, the vessel was taking water through every seam, the crew was constantly at the pumps or baling water out of the cabins with buckets. All the provisions were wet. The seas broke clear over the ship.

On the twenty-sixth, the hurricane winds continued without the least intermission. The ship was continually suffering damages, which had to be repaired as quickly as possible by the exhausted crew. The seas were monstrous, water going through the decks as though they were made of paper. Still the ship was scudding and steadily changing course around the compass. By the twenty-seventh, the weather had improved but the ship persisted in going round and round, veering and scudding before the wind. After all this travel, Captain Finck succeeded in taking an observation and found, to his surprise, that he was not far from port in Mauritius, from which he had set sail before the storm, almost a week earlier, and on the twenty-eighth he made for port there.

From the log kept by Captain Finck and the observations made on other ships caught in the same hurricane, Piddington laid down the track of the storm and the course of the *Charles Heddles.* Now it was clear that the ship had been carried round and round the storm center, at the same time going forward as the storm progressed. Its course at sea looked like a watch spring drawn out—a series of loops extending in an arc from the north to the west of Mauritius. Here was vivid and undeniable proof, from the experience of one ship, that hurricanes over the ocean are progressive whirlwinds, like the storm which Redfield had charted from trees blown down in Connecticut in 1821.

Another fact was quite clear to Piddington and he published it with the hope that all seafaring men would profit by it. He could see now why a ship could be carried hour after hour and day by day before the wind, apparently to great distances, and then be cast ashore near the very place where the ship took to sea.

Inspired by this report of the *Charles Heddles* in the hurricane, Piddington suggested, for the first time in history (1845), that ships be sent out to study hurricanes. He wrote:

"Every man and every set of men who are pursuing the investigation of any great question, are apt to overrate its importance; and perhaps I shall only excite a smile when I say, that the *day will yet come when ships will be sent out to investigate the nature and course of storms and hurricanes,* as they are now sent out to reach the poles or to survey pestilential coasts, or on any other scientific service."

The prediction which Piddington put in italics was eventually verified, though nearly a century later.

"Nothing indeed can more clearly show," Piddington continued, "how this may, with a well appointed and managed vessel be done in perfect safety—performed by mere chance by a fast-sailing colonial brig, manned only as a bullock trader, but capitally officered, and developing for the seaman and meteorologist a view of what we may almost call the *internal* phenomena of winds and waves in a hurricane."

But this was only the beginning. Learning the secrets of the hurricane proved to be far more difficult than Redfield, Reid and Piddington had imagined. The world looked in amazement at the tremendous labors of a few men who collected enormous quantities of reports, interviews, and observations from mariners and tried to put the bits together, but there was a prevailing suspicion that the real facts were locked in the minds of men who had gone to their doom in ships sunk in the centers of these awful storms and the lucky ones who came back had seen only a part of their ultimate terrors. In these days of relatively safe navigation at the middle of the twentieth century, our minds are scarcely able to grasp the seriousness of this scourge of tropical and subtropical seas which destroyed so many ships and drove busy men, working long hours for a living, to such tremendous labors, at night and at odd times, to learn the truth. We may get some light from the stories of desperate sailors who, by some strange fate, were thrown exhausted on the rocks that finally claimed the broken remains of once-proud vessels of trade and war.

3. AT THE BOTTOM OF THE SEA

"Methought I saw a thousand fearful wrecks;
Ten thousand men that fishes gnawed upon:
Wedges of gold, great anchors, heaps of pearl,
Inestimable stones, unvalued jewels,
All scattered in the bottom of the sea."
—Shakespeare

Two hundred years ago, scientists were beginning to chart the winds over the oceans and the currents that thread their way across the surface of deep waters. Until this work was finished, the mariner was almost completely at the mercy of the atmosphere and the sea. He would come to uncharted places where the winds ceased to blow and sailing vessels might be becalmed for weeks. Day after day, the burning sun climbed slowly toward the zenith and while the unbearable heat tortured the crew, descended with agonizing slowness toward the western horizon. At night, relief came under unclouded skies but the stars gave no indication of better fortunes on the morrow.

In these places it seldom rained. Drinking water, as long as it lasted, became putrid, but the crew preserved it as their most precious treasure, drinking a little when they could go no longer without it—holding their noses. The food became so bad that every man who had the courage to eat it wondered if it wouldn't be better to starve. This happened often in the North Atlantic in the days when sailing vessels were carrying horses to the West Indies. If they were becalmed and fresh water ran short, the crews had to throw some or all of the horses overboard. In time this region became known as the "horse latitudes." Because it lay north and northeast of the hurricane belt, a long spell of rainless weather for a sailing ship here could be succeeded suddenly and overwhelmingly by the torrential rains of a tropical storm.

At long intervals, a slight breeze came along, barely enough to extend a small flag, but it gave the ship a little motion and brought hope to the men who were worn out with tugging at the oars. In this circumstance, it might happen that a long, low groundswell would appear. Coming from a great distance, it would raise and then lower the vessel a little in passing. Others would surely follow—low undulations at intervals of four or five to the minute—bringing a warning of a storm beyond the horizon. Here was one of the ironic twists of a sailor's existence. Even while he prayed for water, the atmosphere was about to give it to him in tremendous quantities, both from above and below. At this juncture the master was in a quandary. For the safety of ship and crew, it was vital that he know exactly what to do at the very instant when the first gusty breezes of the coming storm filled the sails.

From the law of storms, the mariner eventually learned —and it was suicide to forget it at a time like this—that if he could look forward from the center of the hurricane, along the line of progress, the most terrible winds and waves would be on his right. Here the raging demons of the tropical blast outdo themselves. The whirling velocity is added to the forward motion, for both in these few harrowing hours have the same direction. All the power of the atmosphere is delivered in this space, where unbelievable gales try to blast their way into the partial vacuum at the center. But the atmosphere is held back from the center by a still greater power, the rotation of the earth on its axis. No shipmaster should ever be caught between these awful forces with the huge bulk of the storm drawing toward him.

Here we find horrors that were never disclosed to the early storm hunters. It is doubtful if any sailing ship or any man aboard survived in this sector of a really great hurricane. But even more dangerous are the deceitful motions of the sea surface, which can trap the mariner and drag his vessel toward the dangerous sector, even while he thinks he is fighting his way out of it.

In those uneasy hours when the groundswell preceded the winds, the master had to watch his barometer and the clouds on the horizon, to get the best estimate of the storm's future course. If it gave signs of coming toward him or passing a little to the west of him, he had to run with the wind as soon as it began, every inch of canvas straining at the creaking masts to get all the headway possible. He would do better than he thought, for the surface of the sea was moving with the winds and his vessel was plowing through the waves while the sea was swirling in the same direction. It was a race for life, and if he was not unlucky, he would find himself behind the storm, sailing rapidly toward better weather.

If he made the wrong choice and tried to go around the center on the east side while the storm moved northward, he might have thought that he was making headway. But the sea surface was carrying him backward while the horrible right sector rushed forward to encompass the ship.

Now we see why Redfield, Reid and Piddington, when they came to a realization of some of these facts in the logs of sailing vessels, were so eager to give the world a law of storms. Their work was only a beginning, for the so-called law is not as simple as they imagined. But some shipmasters took their advice and survived, whereas any other course would have taken them to the bottom of the sea. And untold numbers had gone down in big hurricanes.

Among the logs and letters collected by Redfield and Reid in their work on the law of storms were many which referred to a fierce hurricane in 1780. For more than fifty years it had been talked about as "The Great Hurricane." But the stories didn't all seem to fit together. The storm was said to have been in too many places at too many different times to suit Redfield. When he had finished putting the data from ships' logs on a map in accordance with his law of storms, he saw that there had been three hurricanes at about the same time and that they had been confused and reported as one.

In the year of these big hurricanes there were many warships in the Caribbean region. The American War of Independence had started with bloodshed at Lexington and Bunker Hill in 1775, and by 1780 England was in a state of war with half the world. Her battle fleets controlled most of the seas along the American Coast and roamed the waters in and around the West Indies.

The first of the three hurricanes struck Jamaica on the third of October. Nine English warships, under the command of Sir Peter Parker, went to the bottom. Seven of his vessels were dismasted or severely damaged. From the tenth to the fifteenth of October a second—and even more powerful hurricane—ravaged Barbados and progressively devastated other islands in the Eastern Caribbean. This one has been rated the most terrible hurricane in history by many students of storms. It wreaked awful destruction on the island of St. Lucia, where six thousand persons were crushed in the ruins of demolished buildings. The English fleet in that vicinity disappeared. Neither trees nor houses were left standing on Barbados. Off Martinique, forty ships of a French convoy were sunk and nearly all on board were lost, including four thousand soldiers. On the island itself, nine thousand persons were killed. Most of the vessels in the broad path of the storm as it progressed farther into the Caribbean, including several warships, foundered with all their crews. It drove fifty vessels ashore at Bermuda, on the eighteenth.

Before this terrible storm reached Bermuda another one roared out of the Western Caribbean, crossed western Cuba and passed into the Gulf of Mexico, on October 18. Unaware of the approach of this hurricane, a Spanish fleet of seventy-four warships, under Admiral Solano, sailed from Havana into the Gulf, to attack Pensacola. They were trapped in the eastern section of the Gulf and nineteen ships were lost. The remainder were dispersed, several having thrown their guns overboard to avoid capsizing. Nearly all the others were damaged, many dismasted. The Spanish fleet was no longer a fighting force.

Within three weeks most of the battle fleets in and around the Caribbean had been put out of commission. Both Redfield and Reid were impressed by the power displayed by these hurricanes. In his search of the records, the former succeeded in getting a copy of a letter written by a Lieutenant Archer to his mother in England, giving an account of the first of these terrible storms. The following story is condensed from Archer's letter.

Archer was second in command of an English warship named the *Phoenix*. It was commanded by Sir Hyde Parker. Before the first of these three hurricanes developed, the *Phoenix* had been sent to Pensacola, where the English were in control. Late in September, she sailed to rejoin the remainder of the fleet at Jamaica. On passing Havana harbor, Sir Hyde looked in and was astounded to see Solano's Spanish fleet at anchor. He hurried around Cuba into the Caribbean, to take the news to the British fleet.

At Kingston, Jamaica, the crew of the *Phoenix* found three other men-of-war lying in the harbor and they had a strong party for "kicking up a dust on shore," with dancing until two o'clock every morning. Little did they think of what might be in store for them. Out of the four men-of-war not one was in existence four days later and not a man aboard any of them survived, except a few of the crew of the *Phoenix*. And what is more, the houses where the crews had been so merry were so completely destroyed that scarcely a vestige remained to show where they had stood.

On September 30, the four warships set sail for Port Royal, around the eastern end of Jamaica. At eleven o'clock on the night of October 2, it began to "snuffle," with a "monstrous heavy appearance to the eastward." Sir Hyde sent for Lieutenant Archer.

"What sort of weather have we, Archer?"

"It blows a little and has a very ugly look; if in any other quarter, I should say we were going to have a gale of wind."

They had a very dirty night. At eight in the morning, with close-reefed topsails, the *Phoenix* was fighting a hard blow from the east-northeast, and heavy squalls at times. Archer said he was once in a hurricane in the East Indies and the beginning of it had much the same appearance as this. The crew took in the topsails and were glad they had plenty of sea-room. On Sir Hyde's orders, they secured all the sails with spare gaskets, put good rolling tackles on the yards, squared the booms, saw that the boats were all fast, lashed the guns, double-breeched the lower deckers, got the top-gallant mast down on the deck and, in fact, did everything to make a snug ship.

"And now," Archer wrote, "the poor birds began to suffer from the uproar of the elements and came on board. They turned to the windward like a ship, tack and tack, and dashed themselves down on the deck without attempting to stir till picked up. They would not leave the ship."

The carpenters were placed by the mainmast with broad axes, ready to cut it away to save the ship. Archer found the purser "frightened out of his wits" and two marine officers "white as sheets" from listening to the vibration of the lower deck guns, which were pulling loose and thrashing around. At every roll it seemed that the whole ship's side was going.

At twelve it was blowing a full hurricane. Archer came on deck and found Sir Hyde there. "It blows terribly hard, Archer."

"It does indeed, Sir."

"I don't remember its blowing so hard before," shouted Sir Hyde, striving to get his voice above the roar of the wind. "The ship makes good weather of it on this tack but we must wear her (to turn about by putting the helm up and the stern of the boat to the wind), as the wind has shifted to the south-east and we are fast drawing up on the Coast of Cuba."

"Sir, there is no canvas can stand against it a moment. We may lose three or four of our people in the effort. She'll wear by manning the fire shrouds."

"Well, try it," said Sir Hyde, which was a great condescension for a man of his temperament to accept the advice of a subordinate. It took two hundred men to wear the ship, but when she was turned about, the sea began to run clear across the decks and she had no time to rise from one sea until another lashed into her. Some of the sails had been torn from the masts and the rest began to fly from the yards "through the gaskets like coachwhips."

"To think that the wind could have such force!" Archer shouted into the gale.

"Go down and see what is the matter between decks," ordered Sir Hyde in a lull.

Archer crept below and a marine officer screamed, "We are sinking. The water is up to the bottom of my cot!"

Archer yelled back, "As long as it is not over your mouth, you are well off." He put all spare men to work at the pumps. The *Phoenix* labored heavily, with scarcely any of her above water except the quarter-deck and that seldom.

On returning, Archer found Sir Hyde lashed to a mast. He lashed himself alongside his commander and tried to hear what he was shouting. Afterward, Archer tried to describe this situation in his letter. "If I was to write forever, I could not give you an idea of it. A total darkness above and the sea running in Alps or Peaks of Teneriffe (Mountains is too common an idea); the wind roaring louder than thunder, the ship shaking her sides and groaning."

"Hold fast," shouted Sir Hyde as a big wave crashed into the ship. "That was an ugly sea! We must lower the yards, Archer."

"If we attempt it, Sir, we shall lose them. I wish the mainmast was overboard without carrying anything else along with it."

Another mountainous wave swept the trembling ship. A crewman brought news from the pump room. Water was gaining on the weary pumpers. The ship was almost on her beam-ends. Archer called to Sir Hyde, "Shall we cut the mainmast away?"

"Ay, as fast as you can," said Sir Hyde. But just then a tremendous wave broke right on board, carried everything on deck away and filled the ship with water. The main and mizen masts went, the *Phoenix* righted a little but was in the last struggle of sinking.

As soon as they could shake their heads free of the water, Sir Hyde yelled, "We are gone at last, Archer. Foundered at sea! Farewell, and the Lord have mercy on us!"

Archer felt sorry that he could swim, for he would struggle instinctively and it would take him a quarter hour longer to die than a man who could not. The quarter-deck was full of men praying for mercy. At that moment there was a great thump and a grinding under them.

Archer screamed, "Sir, the ship is ashore. We may save ourselves yet!"

Every stroke of the sea threatened dissolution of the ship's frame. Every wave swept over her as she lay stern ashore.

Sir Hyde cried out, "Keep to the quarter-deck, my lads. When she goes to pieces that is your best chance."

Five men were lost cutting the foremast. The sea seemed to reach for them as it took the mast overboard and they went with it. Everyone expected it would be his turn next. It was awful—the ship grinding and being torn away piece by piece. Mercifully, as if to give the crew another desperate chance, a tremendous wave carried the *Phoenix* among the rocks and she stuck there, though her decks tumbled in.

Archer took off his coat and shoes and prepared to swim, but on second thought he knew it wouldn't do. As second officer, he would have to stay with his commander and see that every man, including the sick and injured, was safely off the ship before he left it. He wrote later that he looked around with a philosophic eye in that moment and was amazed to find that those who had been the most swaggering, swearing bullies in fine weather were now the most pitiful wretches on earth, with death before them.

Finally, Archer helped two sailors off with a line which was made fast to the rocks, and most of those who had survived the storm got ashore alive, including the sick and injured, who were moved from a cabin window by means of a spare topsail-yard.

On shore, Sir Hyde came to Archer so affected that he was scarcely able to make himself understood. "I am happy to see you ashore—but look at our poor *Phoenix*." Weak and worn, the two sat huddled on the shore, silent for a quarter hour, blasted by gale and sea. Archer actually wept. After that, the two officers gathered the men together and rescued some fresh water and provisions from the wreck. They also secured material to make tents. The storm had thrown great quantities of fish into the holes in the rocks and these provided a good meal.

One of the ship's boats was left in fair condition. In two days the carpenters repaired it, and Archer, with four volunteers, set off for Jamaica. They had squally weather and a leaky boat, but by constant baling with two buckets, they arrived at their destination next evening. Eventually, all the remainder of the crew they had left in Cuba were saved except some who died of injuries after getting ashore from the *Phoenix* and a few who got hold of some of the ship's rum and drank themselves to death.

How many times this drama of death and narrow escape may have been repeated in the three great hurricanes of 1780 is not disclosed in the records. But hundreds of ships and many thousands of men were lost. And at that time no one knew the true nature of these great winds. It was not until more than fifty years had passed and Redfield and Reid examined all the reports that these tremendous gales were found to be parts of three separate hurricanes. This ignorance seems strange, for nearly three hundred years had passed since Columbus ran into his first hurricane.

As Reid worked at great length on these old records in logs and letters, he became confident that Redfield was right about the whirling nature of tropical storms. There were ten hurricanes in the West Indies in 1837 and these supplied Reid with a great deal of added information. One of the most exciting was the big hurricane in the middle of August of that year.

This was a vicious storm which was first observed by the Barque *Felicity* in the Atlantic, far east of the Antilles, on August 12, 1837. The chances are that it came from the African Coast, near the Cape Verde Islands, as many of the worst of them do. By the time these faraway disturbances have crossed the Atlantic and approached the West Indies, they are usually major hurricanes, capable of wreaking great destruction. This one was no exception, but its path lay a little farther to the northward than usual and its most furious winds were not felt on land, even on the more northerly islands in the group.

Ships in its path reported winds which appeared to be of a "rotatory" nature when Reid plotted them on maps. On the fifteenth, the storm passed near Turk's Island and on the sixteenth, was being felt on the easternmost Bahamas.

At this stage, the ship *Calypso* became involved in the storm and was unable to escape. The master, a man named Wilkinson, wrote an account to the owners, from which the following is taken:

"During the night the winds increased, and day-light found the vessel under a close-reefed main-topsail, with royal and top-gallant-yards on deck, and prepared for a gale of wind. At 10 A.M. the wind about north-east, the lee-rail under water, and the masts bending like canes. Got a tarpaulin on the main rigging and took the main topsail in. The ship laboring much obliged main and bilge-pumps to be kept constantly going. At 6 P.M. the wind north-west, I should think the latitude would be about 27°, and longitude 77°W. At midnight the wind was west, when a sea took the quarter-boat away.

"At day-dawn, or rather I should have said the time when the day would have dawned, the wind was southwest, and a sea stove the fore-scuttle. All attempts to stop this leak were useless, for when the ship pitched the scuttle was considerably under water. I then had the gaskets and lines cut from the reefed foresail, which blew away; a new fore-topmast-studding-sail was got up and down the fore-rigging, but in a few seconds the bolt-rope only remained; the masts had then to be cut away."

By this time the wind was even more furious and the seas so high none expected the *Calypso* to survive. The master continued his story:

"My chief mate had a small axe in his berth, which he had made very sharp a few days previous. That was immediately procured; and while the men were employed cutting away the mizenmast, the lower yard-arms went in the water. It is human nature to struggle hard for life; so fourteen men and myself got over the rail between the main and mizen rigging as the mast-heads went into the water. The ship was sinking fast. While some men were employed cutting the weather-lanyards of the rigging, some were calling to God for mercy; some were stupified with despair; and two poor fellows, who had gone from the afterhold, over the cargo, to get to the forecastle, to try to stop the leak, were swimming in the ship's hold. In about three minutes after getting on the bends, the weather-lanyards were cut fore and aft, and the mizen, main, and foremasts went one after the other, just as the vessel was going down head foremost.

"The ship hung in this miserable position, as if about to disappear (as shown in the accompanying reconstruction of the scene by an artist who worked under the direction of the master of the *Calypso*) and then by some miracle slowly righted herself.

"On getting on board again, I found the three masts had

gone close off by the deck. The boats were gone, the main hatches stove in, the planks of the deck had started in many places, the water was up to the beams, and the puncheons of rum sending about the hold with great violence. The starboard gunwale was about a foot from the level of the sea, and the larboard about five feet. The sea was breaking over the ship as it would have done over a log. You will, perhaps, say it could not have been worse, and any lives spared to tell the tale. I assure you, Sir, it was worse; and by Divine Providence, every man was suffered to walk from that ship to the quay at Wilmington."

From such accounts the hurricane hunters gathered the facts which led to a better "law of storms" and made life at sea safer for the officers and men who struggled with sails and masts in tropical gales. But it is most likely that the experiences of the crews of those sailing ships that were caught in the worst sectors of fully developed hurricanes in the open sea were never told. It is not probable that any survived the calamitous weather on the right front of the storm center, where the sea, the atmosphere, the rotation of the earth, and the forward motion of the hurricane are combined in a frenzy of destructive power.

In one sense, all of the men who survived these terrors at sea were hurricane hunters. They had to be. Those who lived were the men who were always alert to the first signs in sea and sky, who knew when one of the big storms of the tropics was just beyond the horizon. They were learning and passing the knowledge along to others. By the middle of the nineteenth century, the mariner had a "law of storms" that kept countless ships out of the most dangerous parts of tropical disturbances.

4. STORM WARNINGS

"I am more afraid of a West Indian hurricane than of the entire Spanish Navy."—McKinley

Strangely enough, government weather bureaus were not set up for the purpose of giving warnings of tropical storms. Maybe there was a feeling in the years before radio that nothing could be done for the sailor on the open sea except to teach him the law of storms. And for the landsman the case looked hopeless until the telegraph came in sight. At any rate, most of the men who began to fly into hurricanes during World War II were astonished to find that, up to that time, the prediction of tropical storms had been a kind of side issue.

Although hurricanes are nearly always destructive and other kinds of storms—the "lows" on the weather map—are generally mild, once in a long time one of these others results in a catastrophe. Starting as a low which is spread weakly over a wide area, with cloudy weather, rain or snow, and gentle winds, now and then the exceptional storm suddenly fills newspaper headlines. Gales and winds of hurricane force bring a blizzard, tornado, bad hailstorm, or torrential rain and a damaging flood. If it really is a bad one, it finds its way into the pages of history. In times past, these storms often struck populous districts, while hurricanes, in early centuries, hit on thinly settled islands or coasts.

So far as we know, the worst storm to devastate the British Isles was one of this kind. It was not a tropical cyclone. It was entirely unexpected, as were most of the big gales in England in the old days. Surprise was one of the elements of danger. The weather is seldom fine in the British Isles, over the English Channel or in the North Sea. Gloom, with fog or low-flying clouds, is the rule. Even on the best days, a damp haze hangs everywhere. It is like looking through a dirty window pane. Into this background of gloom many a big storm stole its way eastward from the Atlantic. The record-breaker tore up the docks, wrecked shipping and crumbled buildings in the year 1703.

Houses were ruined and big trees were blown down. Whole fleets were lost and more than nine thousand seamen were drowned. The most violent winds came at night. Startled by the roar of the storm, Queen Anne got out of bed and found a part of the palace roof had been torn away. One prelate, Bishop Kidder, was buried beneath the ruins of his mansion. Awakened by the giant gusts, he put on his dressing gown and made for the door, but a chimney stack crashed through the ceiling and dashed out his brains. His wife was crushed in her bed. After the gales subsided, London and other cities looked like they had been sacked by an enemy. All over the south of England, the lead roofs of churches were rolled up by the wind or blown away in large sheets.

Though other gales almost as bad as this one came in later years, it was more than a century before the storm hunters made much progress. Not long after 1800, several men with an inquiring mind began to get results. Redfield was one, but he studied hurricanes and not the storms of higher latitudes, such as the one which devasted the British Isles.

Shortly after 1800, there were signs of the coming of faster means of travel and communications and they were destined to be a vital factor in weather forecasting. In 1816 a "hobby-horse" with wheels was displayed in Paris by an inventor named Niepice. It was propelled by a man or two sitting on it and pushing on the ground. Even with two men pushing, it went no faster than a man could walk. But strong claims were made about its possibilities. At about the same time, several men were working on devices like the telegraph.

Whether it was this trend or not, something aroused the intense curiosity of a young professor, William Heinrich Brandes, of the University of Breslau, in Germany. He began a study in 1816, to see if the weather moved from place to place and if it would be possible to send predictions ahead by means then available. Everybody at that time knew that storms moved but it was the general belief that ordinary changes in the weather didn't go anywhere. Brandes collected newspapers from many places and searched them for remarks about the weather, which he put on maps. Here he was amazed to see that all kinds of weather seemed to be constantly in motion, quite generally from west to east. But the newspaper reports were rather poor for his purposes and he couldn't be too sure about the rate of travel.

Brandes knew that the French had set up weather stations and collected observations for maps as early as 1780, but the terrible French Revolution had brought an end to this work and the data were lying in disuse. After some delay, he obtained copies of the observations for 1783 and put them on maps. Sure enough, after he had drawn many daily maps, he saw clearly how the weather moved just as he had suspected it did from the newspaper reports. But at the same time he saw that it was hopeless. The weather moved so rapidly that there was no way of sending the reports ahead fast enough for making predictions of what was coming. The quickest way of sending the reports ahead was by horse or a good man on foot, and the weather would easily outrun them. In 1820, Brandes wrote an article about weather maps for publication and then put his maps and newspapers in the trash. But in time his idea got around the world and as the years passed more and more scientists began drawing maps and trying to predict the weather. And so it came about that the government weather services in different parts of the world were set up to predict storms of higher latitudes rather than hurricanes.

Redfield was mapping storms after 1830, but he was not trying to make weather forecasts. He wanted only to learn about hurricanes in order to give the mariner a law of storms by which he could judge the weather for himself. Nobody worried about the landlubber. It was the idea in those days that a man on land could get his weather out of an almanac or by watching the signs of the winds, clouds, birds, stars, or the rise and fall of the barometer. Scientists who believed that it would be possible to predict the ordinary changes in the weather were decidedly in the minority. One of these was James Pollard Espy, who became known as the "Old Storm King" of America.

James Espy was born in Pennsylvania, in the vicinity of Harrisburg, but his father moved the family to Kentucky while James was an infant. It has been said in biographies of Espy that the boy had no education and was seventeen years old before he learned to read, but this was denied by relatives who survived him. It seems that the elder Espy soon went to the Miami Valley in Ohio, to get established in business, and left James with an older sister in Kentucky. At eighteen James registered at Transylvania University, in Lexington, where he was much interested in science. In any event, at various times he was a schoolteacher in Ohio, Maryland, and Pennsylvania, until he became fully occupied in the study of weather.

In 1820, Espy joined the Franklin Institute, in Philadelphia, to teach languages and work on the weather. In an amazingly short time, he became an authority on meteorology. He was a pleasant, easygoing man, but very persistent in two matters. First, he was determined to have a government bureau established to predict storms; and, second, he disagreed with Redfield in the latter's whirlwind theory of hurricanes. At times the two carried on a violent controversy in the press. Espy argued that the winds blow directly toward the center of a storm or toward a line through the center. He was right with respect to storms of middle and higher latitudes, as everybody knows today. He anticipated the modern idea of fronts, and he and other scientists of his day sometimes referred to these lines as "like a line of battle." In a way, Redfield also was right, for the typical hurricane in the tropics has no fronts.

In his efforts to set up a government weather bureau, Espy was successful in a small way. In 1842, he was appointed by Congress for five years as "Meteorologist to the U. S. Government" and assigned to the Surgeon General, where he worked for five years. This rather strange appointment was due to the fact that the Surgeon General had been taking weather observations at Army posts since 1819 and had much data for study.

In the meantime, Espy had visited England and France, where he was received with honor by renowned scientific associations. On returning to the United States, he published a book, *The Philosophy of Storms,* in 1841. His weather maps and storm reports were now famous and by this time he was widely known as the "Old Storm King." When his term as "Meteorologist to the U. S. Government" expired, he secured an appointment as meteorogolist under the Secretary of the Navy, to work with the Smithsonian Institution, where he made an annual report to the Navy until 1852.

During these years, Espy was continually after Congress to do more about storm hunting. In Washington, he earned the title of the "Half Baked Storm Hunter" and in Congress he was known as the "Old Storm Breeder." In 1842 he was granted hearings and members of an appropriation committee said that he was a "monomaniac" and his "organ of self-esteem was swollen to the size of a goiter." They told him that they were not impressed just because "the French had indorsed all his crack-brained schemes." Espy kept insisting for several years and was looked upon as a nuisance in Congress until he died in 1860, having had very little success in getting the government to do anything about it, except to give him an appointment to study the weather himself.

As it finally worked out, Congress in 1870 established a weather service, to study storms on the Great Lakes and the seacoasts of the United States. This proved to be such a tough job that, for the time being, the hurricane work, which had been neglected during and after the War between the States, was dropped into second place.

The disturbances that kept the government service busy after 1870 are those that begin in higher latitudes and move generally from west to east—the lows of the weather map—called extratropical to distinguish them from hurricanes and other tropical storms. If they were as regular in their shapes and movements as the tropical variety, the forecasting job would be much easier. But the extratropical kind takes odd forms, elongated or in the shape of a trough, sometimes with two or more centers. Their movements are irregular. Rarely does one of them become extremely violent, but there is al-

The English warship Egmont *in the "Great Hurricane" of 1780.*

The Calypso *in the big Atlantic hurricane of 1837, showing the crew climbing over the rail as the mastheads go into the water.*

USWB—Miami Herald Staff Photo

A tremendous wave breaks against the distant seawall on Florida coast at the height of a hurricane.

Official U. S. Navy Photograph

Typhoon buckles the flight deck of the aircraft carrier Bennington *and drapes it over the bow.*

Winds of hurricane drive pine board (10 feet by 1 inch by 3 inches) through the tough trunk of a palm tree in Puerto Rico, September 13, 1928.

Official U. S. Navy Photograph

Looking down from plane at the surface of the sea with winds of 15 knots (17 miles an hour).

Official U. S. Navy Photograph

Sea surface with winds of 40 knots (46 miles an hour).

Official U. S. Navy Photograph

Sea surface with winds of 75 knots (86 miles an hour).

Official U. S. Navy Photograph

Sea surface with winds of 120 knots (138 miles an hour). Tops of big waves are torn off and carried away in a white boiling sheet.

Superfortress B-29 used by Air Force for hurricane hunting.

Official U. S. Navy Photograph

Neptune P2V-3W used by Navy for hurricane hunting.

Official U. S. Navy Photograph

Navy crew of hurricane hunters.

Air Force Photo

Air Force crew being briefed by weather officer before flight into hurricane.

Official U. S. Navy Photograph

Conditions at birth of Caribbean Charlie in 1951.

In the foreground, part of a spiral squall band, an "arm of the octopus."

Photographed by McClellan Air Force Base

Through Plexiglas nose, weather officer sees white caps on sea 1,500 feet below.

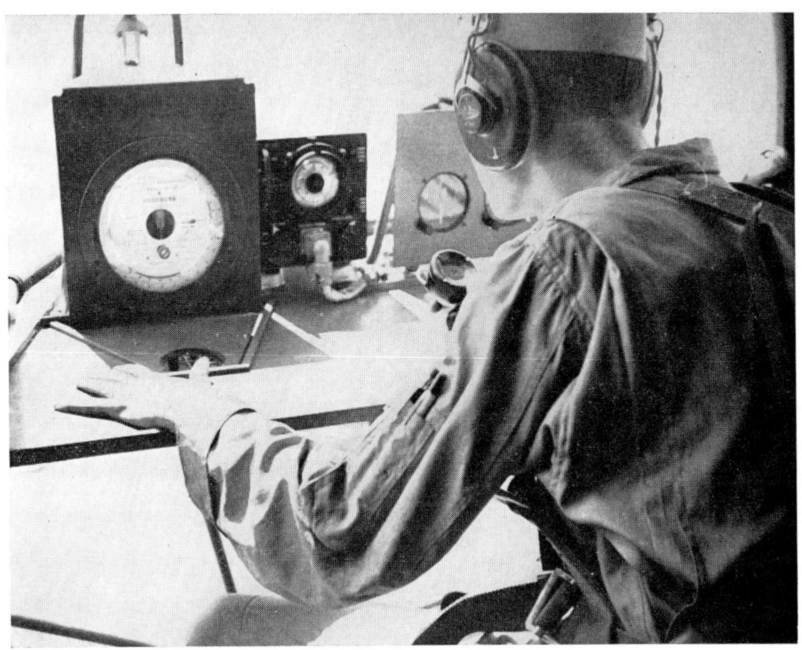

Official U. S. Navy Photograph

Navy aerologist at his station in nose of aircraft on hurricane mission.

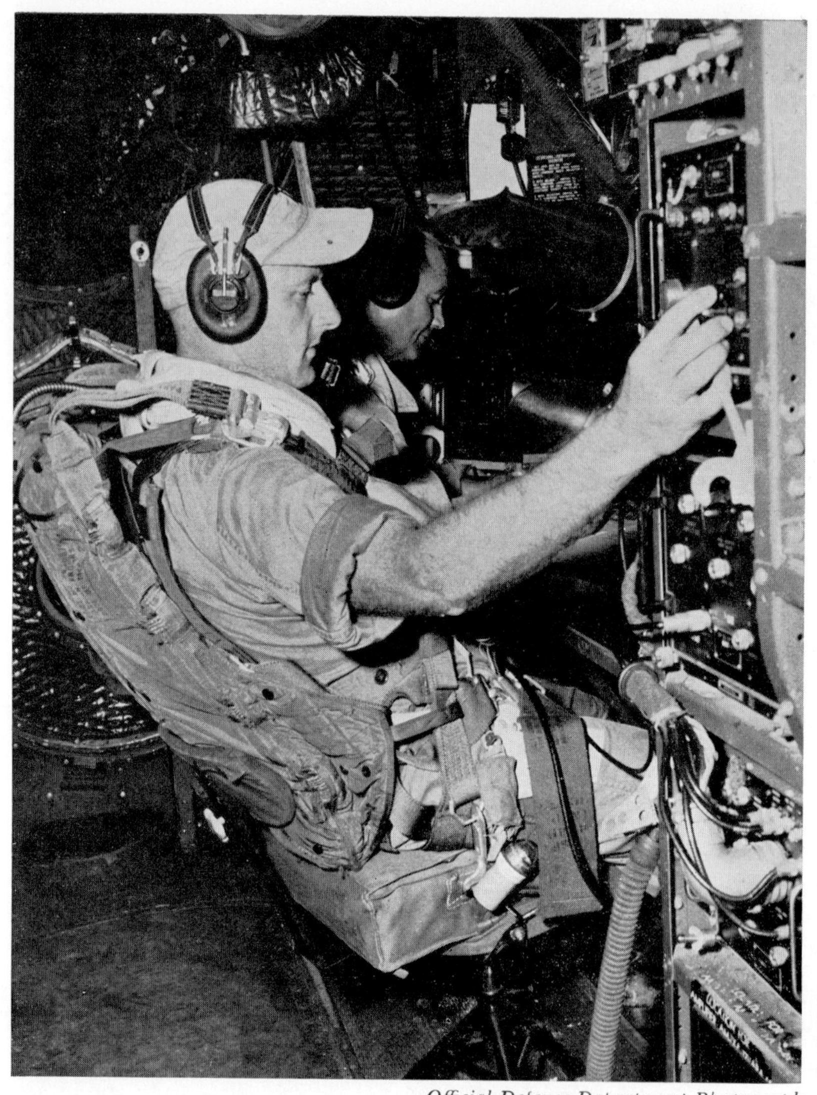

Official Defense Department Photograph

Radar operator in foreground; navigator in background.

Official Defense Department Photograph

Maintenance crew goes to work on B-29 after return from hurricane mission.

USWB—Miami Daily News

City docks at Miami after passage of Kappler's Hurricane in September, 1945.

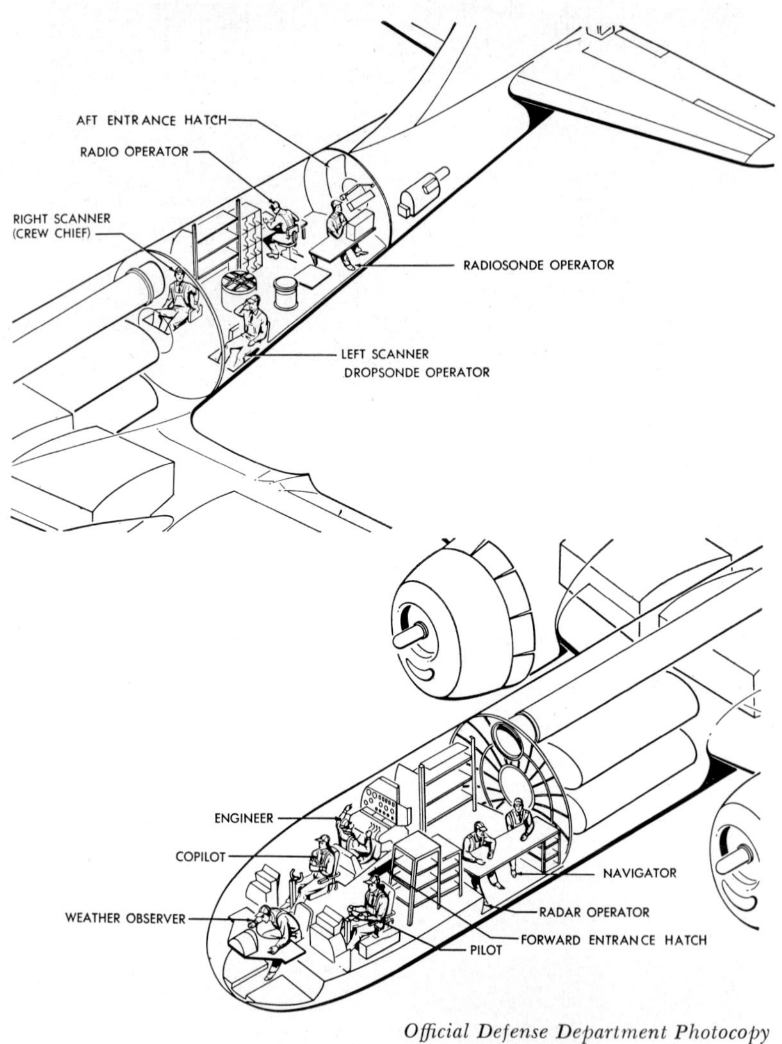

Official Defense Department Photocopy

Positions of crew members in B-29 on hurricane mission.

Part of scope showing typhoon by radar. Eye is above center at right with spiral bands showing. Radar is located at center of picture with surrounding clouds showing as dense white mass due to heavy nearby echo return. Echo from opposite side of typhoon is faint.

Official U. S. Navy Photograph

Looking down into the eye of Hurricane Edna (foreground) on September 7, 1954.

U. S. Air Force Photo

Looking down at the central region of Typhoon Marge in 1951.

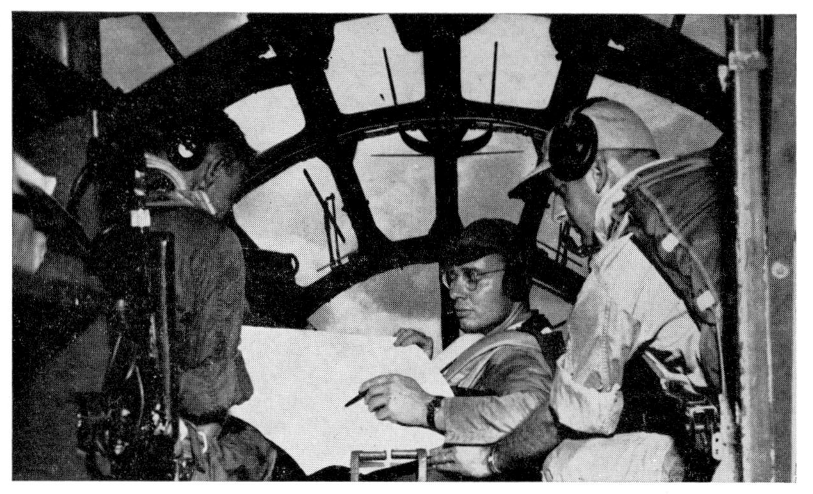

Weather officer in nose of aircraft talking to pilot (left) and radar operator.

Official Photo U. S. A. F.

The engineer in a B-29 on hurricane reconnaissance.

Official Defense Department Photograph *Official Photo U. S. A. F.*

The two scanners ready to signal engine trouble the instant it shows up.

The new plane (B-50) to be used by the Air Force for hurricane reconnaissance.

ways danger of it and so the forecasters must always be on the alert.

Some of the most dangerous of the extratropical storms begin as small companions or secondary centers of huge disturbances, generally on the south side, where they grow rapidly in fury and merge with the original cyclones to produce winds of tremendous destructive power. This often happens in the so-called "windy corners" of the world. One of these, and a good example, is Cape Hatteras, on the eastern coast of North Carolina. It is a sort of way station for both the tropical and extratropical varieties. Hurricanes heading northwestward from the Caribbean and curving to follow the coastline, sweep over the Cape, which juts into the ocean at the point where the northward-moving storms still retain great force. In winter, big extratropical cyclones passing eastward across the region of the Great Lakes tend to produce small companions or secondaries in the southeastern states, and some of them develop gales of hurricane force by the time they reach Hatteras. Here the cold air masses of the continent, guided by storm winds, are thrown against the warm, moist air from the Gulf Stream. In the reaction, there are towering seas and hazardous gales that are well known to seamen.

As these big storms roar past Cape Hatteras, the winds shift to northwest and the sky clears, unless you happen to be on shipboard and the tops of big waves are being torn off by the wind and thrown into the air, to pass overhead in streaks or splatter on the decks. In the days of the sailing ship, the master was not surprised when he got into trouble in the area between Bermuda and Hatteras. Here many merchantmen from far places passed, en route to or from New York or other Atlantic ports. Slowed by cross seas and dirty weather hatched over the Gulf Stream, they were soon reduced to storm stay-sails. As the gales mounted, the crews could see other ships rising on the billows in one instant before slithering into a great trough where, in the next instant, they could see nothing but jagged peaks of water and a welter of foam. On the Hatteras side, especially, the master could get into a rendezvous with death, for he often had only two choices. He could run full tilt toward the west and try to get around the front of a hurricane moving northward, but this maneuver would take him toward Hatteras, where he might find company in the wrecks of countless other ships that had failed in the effort and had been thrown against the coast. The other choice was no better. He could make such progress as was possible toward the east and hope that he would not be caught in the dangerous sector of the oncoming hurricane, a course which more likely than not would lead to disaster.

As has been noted, however, it was the tragic losses caused by extratropical cyclones that induced governments to take over the job of hunting storms and issuing warnings. In France, the first country to take positive action, the immediate cause was the catastrophe which struck the allied fleets in the harbor at Balaclava in 1854, during the Crimean War. Ships of England and France were caught in this desperate position because of jealousies and hatreds which have abounded in Europe for centuries. In this case, the Tsar of Russia seized a pretext to try to gain control of a part of Turkey. This was not unexpected. Russia always has looked with covetous eyes at the Bosporus and the Dardanelles, which lead through the Black Sea to the Mediterranean. On this score Europe is perpetually uneasy. France and England, who had been enemies, now joined forces and planned a campaign against Russia.

It was July, 1853, when the Tsar, Nicholas I, mobilized his armies. As his first overt act, he occupied the part of Turkey which lay north of the Danube River. Soon afterward, the Russian fleet destroyed a Turkish squadron in the Black Sea. Now the Tsar became more cautious because of the threat of action by England and France, and especially because of indications that Russia's ally, Austria, would desert her. The Tsar took no further action. Now it required a long time in those days to get a campaign under way, and it was a whole year later, July, 1854, when the allies were ready to start the invasion of the Crimean Peninsula. Meanwhile, Russia had withdrawn her troops from Turkey and there was no real cause for conflict. But tempers had flared, the vast machinery of war had been put in motion, and the allies drew stubbornly nearer to disaster. They knew quite well that the time might be too short to finish the campaign before the bitterly cold weather of the Russian winter would creep out over the Peninsula. In fact, the Tsar had said that his best generals were January and February, and that remark should have carried ample warning.

Actually, the allied attack began in September, 1854. The British had taken possession of the harbor at Balaclava and, in the beginning, the invasion seemed to promise success. But in October the heroic but ill-fated "Charge of the Light Brigade," made immortal by Tennyson, marked the turning point. It was clear then that the campaign would have to be resumed in the spring of 1855. By November, cold weather had arrived, land action had ceased, and the allies were faced with the problem they had hoped so earnestly to avoid —that of keeping their fighting forces intact during winter in a hostile climate.

To understand the dire predicament of the allies when the big storm struck, it is important to note that the harbor at Balaclava had proved to be too small for a supply base. Many ships had to be anchored outside and there was delay and confusion in moving in and out of the harbor. Not only was there a difficult supply problem but the sick and wounded were being transported across the Black Sea to Scutari, near Constantinople, where hospital conditions were abominable. By October, the plight of the army had become a scandal in England. Florence Nightingale was sent to Scutari with authority over all the nurses and a guarantee of co-operation from the medical staff. She arrived on November 4. The remainder of her story is well known as one of the bright pages of history.

Now the stage was set for catastrophe. An obscure winter storm blew its way across Europe without anything happening until its southern center crossed the Black Sea, on November 14. Suddenly, as secondaries often do, it came to life. There was rain turning to snow as the disturbance burst forth in gales of hurricane force. The congestion grew while the signs of the storm intensified. The ghostly mountains around Balaclava disappeared in the gloom, the near-by shore lines next were blotted out, and impenetrable darkness settled down on the shuddering and grinding of the battered remnants of the helpless fleet. Wreckage was strewn along the coast and around the harbor. All the men-of-war survived, although damaged, but nearly all of the vessels with essential stores were lost.

Misery, disease, and horror followed during the bitter winter. The death rate in the hospitals reached forty-two per cent in February. Meanwhile, in France, Napoleon III received news of the terrible gales at Balaclava and brooded over the catastrophe. He determined to learn where this deadly storm had originated, the path it followed, and to set up a plan for tracking and predicting others of its kind in the future. And so he called in the famous astronomer Leverrier and asked him to carry out the investigation.

Urbain Leverrier, then forty-three years old, was known throughout the world as the discoverer of the planet Neptune, in 1846. He knew of the works of Redfield and Reid on hurricanes and by 1854 had noted the efforts of other Americans and Britishers to track extratropical storms. With their ideas in mind, he called on scientists in all European countries to send him observations of the weather on the days from November 12 to 16, preceding and following the day of the disaster at Balaclava. Information moved slowly between countries in those days and, though many scientists co-operated, it was February, 1855, before Leverrier had gathered the data he needed. In developing his plan, he was encouraged by the invention and spread of the electric telegraph in the United States, and he hoped that the extension of lines in Europe would provide fast-moving messages for his purpose.

Before the end of February, Leverrier handed his report to Napoleon III and recommended that a system of weather messages and of issuing warnings be established at once. The Emperor approved this within twenty-four hours. Soon the French government was mapping the weather and looking for storms. The British followed suit. Already Joseph Henry, in the Smithsonian Institution in Washington, was trying a similar plan, but it was not until February, 1870, that the Congress of the United States appropriated funds and established a government weather service in the Signal Corps of the Army.

The immediate reason for this legislation in the United States was similar to that in France. At that time there was a rapidly growing commerce on the Great Lakes, but storm disasters were all too frequent. In 1869, nearly two thousand vessels were beached or sunk by gales on the Lakes. On the seacoasts, the situation was almost equally bad. The new service was soon in operation. The first storm warning by the United States government was sent out in November, 1870.

During the next twenty years, blizzards, hail storms, tornadoes and sudden wind storms of other kinds gave the new weather service a great deal of trouble. They brought a vivid realization of the great variety of surprises that lay in wait for the storm hunters. No sooner had they found rules for the issuance of warnings than a new kind of peril came along. The service had been in the Signal Corps of the Army, but in 1891 it was turned over to the Department of Agriculture because of its value to the farmers. The desperate struggle against storms continued, with many experienced weathermen feeling very discouraged about the whole business. And then on February 15, 1898, the Battleship *Maine* was sunk in Havana Harbor and war with Spain loomed on the horizon.

On April 25, the United States declared war. The Spanish fleet left the Cape Verde Islands for Cuba and American warships departed for the West Indies, to prepare the way for the movement of troops for the coming campaign in Cuba. It was June 29, however, before the transports arrived at Santiago, carrying seventeen thousand officers and men to support the United States fleet. By that time, the commanders on both sides had begun to worry about storms, for the first hurricanes had appeared as early as June in some years, bringing destructive winds and torrential rains to some parts of Cuba and the surrounding area.

Willis Moore was Chief of the Weather Bureau. He had been a sergeant in the Signal Corps, transferred when the service was put in the Department of Agriculture. He knew very well the difficulties of tracking storms and especially in the West Indies, where only scattered weather reports could be obtained by cable from some of the islands. A bad hurricane could easily sneak up on the American forces through the broad waters of the Caribbean, a predicament likely to arise if the Weather Bureau depended on cable messages from native observers.

Moore carried his worries to James Wilson, Secretary of Agriculture, who decided that they should go to the President. At the White House, they soon had an audience with McKinley, and Wilson presented the case. Moore had maps, charts and data on hurricanes and the disasters they had caused in the West Indies. Also, he had sketched a plan for a cordon of storm hunters on islands around the Caribbean, to protect the American fleet. He said that armadas had been defeated, not by the enemy, but by the weather. He thought it probable that as many warships had been sent to the bottom by storms as by the fire of the enemy. The President listened respectfully at first, then with impatience at the lengthy discussion. He had made up his mind. Interrupting Moore, he got up, sat on the corner of his desk and declared:

"Wilson, I am more afraid of a West Indian hurricane than I am of the entire Spanish Navy. Get this service started at the earliest possible moment."

Moore ventured to say, "Yes, indeed, Mr. President, but the Weather Bureau will need the authority of Congress to organize a weather service on foreign soil."

The President told Wilson: "Report to Chairman Cannon of the Appropriations Committee at once. They are preparing a bill to give me all necessary powers to conduct the war and this authority can be included."

It was soon done. As a part of the plan, a fast cruiser was stationed at Key West, to carry the news to the fleet immediately, in case the Weather Bureau predicted a hurricane. In that event, the fleet might have abandoned the blockade, to get sea room and avoid the center of the storm.

With this authority, the Weather Bureau moved swiftly to station men and equipment on the islands. Letters had to be written to European countries for permission to send observers into their possessions. But although the bill containing the authority only passed Congress on July 7, observers arrived as follows: July 21—Kingston, Santiago, Trinidad, San Domingo, St. Thomas; August 11—Barranquilla; August 12—Barbados; August 18—St. Kitts; August 29—Panama.

Land fighting continued in the West Indies until August 12, but the Spanish fleet was destroyed on the morning of July 3. They made a desperate effort to escape from the harbor at Santiago, were shelled by American warships, and all were disabled or beached. Up to that time there had been no tropical disturbances in the region. A small one hit near Tampa on August 3. Another small but vicious hurricane swept the coast of Georgia on August 31. The first big one of the 1898 season raked Barbados, St. Vincent and St. Lucia on September 10 to 11, and disappeared east of the Bahamas.

The stations set up by the storm hunters in 1898 formed the backbone of the hurricane warning service which exists today as a greatly improved system, including squadrons of aircraft that fly into tropical storms to obtain essential data for the forecasters. Before storm hunting could be operated on a practical basis, however, it was necessary to find new means of communication. Dependence on messages by cable from scattered islands was not good enough.

5. RADIO HELPS—THEN HINDERS

>>>

"Make it clear that I would veto the bill again."—F. D. R.

In the 1930's there was a strange turn of affairs in hurricane hunting. It had long been the purpose to keep ships out of trouble, first by giving the mariner a law of storms and then by sending warnings by radio. One morning in August, 1932, an indignant citizen came into a Weather Bureau office on the Gulf Coast and wanted to know where the hurricane was. The weatherman told him that there were no ship reports in the area but the center seemed to be somewhere in the central Gulf.

"What's the matter with the radio reports from boats?" he asked.

"Because of the warnings we issued yesterday, all the ships got out of the area and apparently there are no ships close enough this morning to do any good," the weatherman explained.

"Say, what kind of a deal is this?" demanded the citizen. "The only way we can tell where the center is located is to get radio reports from boats out there and you fellows chase all the boats away from the storm."

"Well, that's our business," replied the Weather Bureau man in astonishment. "We are required by law to give warnings to shipping."

"I don't see it. I'm going to write to my Congressman and to the White House, if necessary, to get this straightened out. What we ought to do is send boats out there to give reports when we need them," was the final declaration by the citizen who had one time been a shipmaster himself. And he did write to Congress and the White House. Others joined him. The argument over legislation began.

Long before the use of radio on shipboard, the location, intensity, and movement of hurricanes over the Atlantic, Caribbean, and Gulf, and along the coasts and between the islands in the West Indies had been judged by careful observations of the wind, sea and sky. In the latter part of the nineteenth century, the storm hunters had become quite expert at it. Among the best were the Jesuits in the West Indies and in the Far East. They watched the high clouds moving out in advance of the tropical storm, the sea swells that are stirred up by the big winds and travel rapidly ahead, and, finally, as the storm center drew near, they studied the winds in the outer edges when they began to be felt locally. One of the pioneers in this work in the West Indies was Father Benito Viñes, at Havana. He began giving out warnings as early as 1875 and by the end of the century was an authority on the precursory signs of hurricanes, both for land observers and for men on shipboard. By that time many of the Weather Bureau men along the coasts had become experts and, after the Spanish War, they began work on the islands in the West Indies.

Observations from the islands came in by cable and from the American coasts they came by telegraph. In some areas this information served very well, but far from land—in the open Atlantic, Caribbean, or Gulf—there was not much to go on. Along the Gulf and Atlantic coasts, the last resort before putting up the red flags with black centers was the experienced observer who had an unobstructed view of the open sea. Even with the best of such reports, there was always a question as to whether it was a big storm with its center far out or a small storm with its center close by. This fact, plus the rate of forward motion of the storm, could make a vital difference. A big, slow-moving storm gave plenty of warning but a small, fast-moving one brought destructive winds and tides almost as soon as the warnings could be sent out and the flags hoisted.

Aside from these indications, the storm hunters depended heavily on the behavior of tropical storms in different parts of the season. They had average tracks by months, showing how storms had moved both in direction and speed, and much other information on their normal behavior. But all too often hurricanes took an erratic course, and now and then the center of a big one described a loop or a track shaped like a hairpin. A few of the storm hunters thought that some upper air movement—a "steering current"—controlled the hurricane's path. The most obvious influence of this kind is the general air circulation over the Atlantic—the large anticyclone nearly always centered over the ocean near the Azores but often extending westward to Bermuda or even to the American mainland.

In the central regions of the Atlantic High, the modern sailor, unlike his predecessor in the sailing ship, is delighted by calms or gentle breezes and fair weather. On its northern edge, storms pass from America to Europe, stirring the northern regions of the ocean. On its southern edge, we find the trade winds reaching down into the tropics and turning westward across the West Indies and the Bahamas. A chart of these prevailing winds gives a fairly good indication of the ocean currents. Some of the surface waters are cold, some warm. And where they wander through the tropics as equatorial currents or counter-currents, they are hot and, other things being favorable, we find a birthplace of storms. In some other tropical regions, the waters are cold and no hurricanes form there.

Near the equator, the earth is girdled by a belt of heat, calms, oppressive humidity, and persistent showers. This belt is called the "doldrums." The trade winds of the Northern Hemisphere reach to its northern edge, while the trades below the equator brush its southern margin. Tropical storms form now and then in and along the doldrum belt at certain seasons—just why, no one knows, for there are hundreds of days when everything seems right for a cyclone but nothing happens except showers and the miserable sultriness of the torrid atmosphere.

Stripped to his waist, the sailor sits on his bunk at night without the slightest exertion while perspiration descends in rivulets from his head and shoulders. Nothing seems capable of making any appreciable change in this monotonous regime. But eight or ten times a year on the Atlantic, in summer or autumn, a storm rears its head in this oppressive atmosphere. Its winds turn against the motions of the hands of a clock, seemingly geared to the edges of the vast, fair-weather whirlwind centered in mid-ocean. Around the southern and western margins of this great whirl the storm moves majestically, gaining in power which it takes in some manner from the heat and humidity—a power which would drain the energies of a thousand atom bombs. The crowning clouds push to enormous heights and deploy ahead of the monster—a foreboding of destruction in its path. Here is one of the great mysteries of the sea. Its heated surface lets loose great quantities of moisture which somehow feed the monster—that we know—but what sets it off is almost as much of a mystery as it was in the time of Columbus.

Until lately, the investigators trying to study the hurricane in motion across the earth were as handicapped as if they had been stricken blind and dumb when its great cloud shield enveloped them. The darkening scud and rain shut off all view of the upper regions by day and left them in utter darkness by night. No word came from ships caught in its inward tentacles until long afterward, when the survivors had come into port. Balloons tracing its winds disappeared in the clouds and were carried away. A method of following them above the clouds would have helped in the understanding of the upper regions in the same way that reports from sailing ships had helped in the study of the surface winds. This was the situation at the end of the Spanish War. But a new era was opening.

As the century came to a close, Marconi was getting ready to span the far reaches of the Atlantic with his wireless apparatus. Already the miracle of the telephone carrying the human voice by wire had become a practical reality, with more than a million subscribers in the United States, but it was not destined to be used across the ocean for many years. Even that accomplishment would not have afforded much help to the storm hunters. They had tried transoceanic messages for weather reporting when submarine cables were laid across the Atlantic. Some weathermen thought at first that it would be possible to pick up reports of storms on the American Coast and, allowing a certain number of days for them to cross the Atlantic, to predict their arrival in Europe. This failed to work, for many storms die or merge with others en route, and so many new disturbances are born in mid-Atlantic that it is necesssary to have reports every day from all parts of the ocean to tell when storms are likely to approach European shores.

In 1900, Marconi was building a long distance transmitting station in England, and readable signals had been sent over a span of two hundred miles. No one then could foresee the strange roles that this remarkable invention would play in storm hunting but it was obvious that messages could be sent across long distances between ships at sea and from ship to shore. Already wireless had been used successfully between British war vessels on maneuvers. Actually, it was destined to be a powerful ally of the men who searched for hurricanes and reported their progress, but eventually this trend reversed itself and radio was the cause of tropical storms being found and then lost again in critical circumstances.

The spread of wireless across the oceans began while the American people still had vividly in mind the most terrible hurricane disaster in the history of the United States. The nation had been shocked by news of a "tidal wave" which had virtually destroyed Galveston, Texas, on the night of September 8, 1900, and killed more than six thousand of its citizens. Really it was not a tidal wave but a West Indian hurricane of almost irresistible force which had raised the tide to heights never known before and then topped it with an enormous storm wave as the center struck the low-lying island.

There was good reason to expect a disaster of this kind. A number of bad hurricanes had hit Galveston in the nineteenth century. The first of which we have any reliable record struck the island in 1818, when it was nothing more than a rendezvous for pirates, principally the notorious Jean Lafitte. It is known that he was in full possession there in 1817, and it was rumored that he and his pirate crews were caught in the hurricane of 1818 and had four of their vessels sunk or driven on shore.

All along the Texas Coast, the inhabitants always have worried about hurricanes and they have plenty of reason. Whole settlements have been destroyed by wind and wave.

One case deserves special mention. After the middle of the century, there had been a thriving town named Indianola in the coastal region southwest of Galveston. The town gave promise then of being the principal competitor of the island city for the commerce of the State of Texas. But in September, 1875, a West Indian hurricane took a slow westward course through the Caribbean Sea and the Gulf of Mexico, and struck the Coast near Indianola. Vicious winds prostrated the buildings while enormous waves swept through the streets, drowning a large share of the population.

Courageous citizens rebuilt the town and for more than ten years it prospered. Then in August, 1886, a bigger hurricane ravaged the town and the countryside and literally wiped the place out of existence. The survivors deserted the site and after a few days nothing was left to mark the spot except sand, bushes and the wrecks of houses and carriages, a litter of personal property, and a great many dead animals. After the hurricane of 1875, the Signal Corps had established a weather station at Indianola, and in the storm of 1886 the building fell in, overturning a lamp in the office and setting fire to the fallen timbers. The observer tried to escape but was drowned in the street.

Both of these hurricanes caused much damage at Galveston, for the island was caught in the dangerous sector on the right of the center in both cases. And it was natural that when, on September 8, 1900, the winds began to increase and the tide rose above the ordinary marks at Galveston, the citizens became alarmed, expecting a repetition of the big blows of 1875 and 1886, which were still being mentioned in August and September every year when the Gulf became rough and gusty northeast winds tugged at the palm trees and oleanders.

But on September 8 the wind kept on rising and the tide crept above any previous records. The weather observers feared the worst and dispatched a telegram to Washington, telling about the heavy storm swells flooding the lower parts of the city and adding, "Such high water with opposing winds never seen before." It was not altogether unexpected. Beginning on September 4, the hurricane had been tracked across Cuba and into the Gulf toward the Texas Coast, but this rise of the sea was more than the observers had bargained for.

By noon, the wind and sea were much worse, the fall of the barometer was ominous, and the Signal Corps observers, two brothers named Isaac and Joe Cline, took turns going out to the beach and reporting to Washington. At 4:00 P.M., all communications failed. Isaac found the water waist deep around his home and the wreckage of beach homes battered by waves was flying through the streets. At 6:30, Joe, who had come to the south end of the city to view the Gulf, joined his brother and found the water neck deep in the streets and roofs of houses and timbers flying overhead after being tossed into the air by giant waves. As the peril grew, fifty neighbors gathered for refuge in the Cline home because it was stronger than others in that part of the city.

At 6:30, in the weather office, one of the assistant observers, Joe Blagden, looked first at the steep downward curve on the recording barometer and then noted that the wind register had failed as the gale rose to one hundred miles an hour. To repair the gauge, he climbed to the roof and crawled out, holding on tightly in the gusts and edging forward in the lulls. Reaching the instrument support, he saw that the wind gauge had been blown away, so he crawled down from the roof, after taking one brief, horrified look over the stricken city.

There was no longer any island—just buildings protruding from the Gulf, with the mainland miles away. Down the street filled with surging water, the spire of a church bent in the wind and then let go as the tower collapsed. The side of a brick building crumbled. As each terrible gust held sway for a few moments, the air was full of debris. The top story of a brick building was sheared off. The scene was like that caused by the destructive blasts at the center of a tornado but, instead of the minute or two of the twister, it lasted for hours. Darkness, under low racing storm clouds, swiftly closed over the city in the deafening roar of giant winds and the crash of broken buildings. The frightened observers saw that the right front sector of the hurricane was bearing down on the island.

Out at the beach, block after block of houses, high-raised to keep them above the tide marks of previous storms, had been swept into the center of the city and were being used as battering rams to destroy succeeding blocks, until a great pile of wreckage held against the mountainous waves. After an hour or two that seemed like an eternity, the hurricane center began crossing the western end of the island, and the city on the eastern end was swept by enormous seas which brought the water level to twenty feet behind the dam of wrecked houses. Everything floated, many frame buildings, or what was left of them, being carried out into the Gulf.

The Cline house disintegrated and more than thirty people in it drowned, among them Isaac's wife. The others drifted on wreckage, rising and falling with huge waves and trying desperately to hold timbers between them and the wind, to ward off flying boards, slate, and shingles. One woman, seeing her house was giving way to the wind and going down in the water, fastened her baby to the roof by hammering a big nail through one of his wrists. He survived. How many drowned or were killed in that awful night was never known. The estimates finally rose above six thousand.

Doubt about the number was due to the presence of many summer visitors at the beaches and, besides, there was no accurate check on the missing, partly because the cemetery was washed out and the recently buried dead were confused with the bodies of storm victims. The aftermath was horrible beyond description.

Galveston had been on the right edge of the hurricane center. If the city had been equally close to the center on the left side, the destruction of wind and waves would have been bad, but nothing like that actually experienced. On the left side—that is, left when looking forward along the line of progress—the tide would have fallen rapidly as the center passed and the gales would have lacked the peak velocities so damaging to brick buildings and other structures which had withstood previous hurricanes. Here was a sharp challenge to the storm hunters. To tell in advance how devastating the hurricane might be, they would have to be able to predict its path with sufficient accuracy to say with some assurance whether the center would pass to the left or right of a coastal city.

This case shows how hard it was to make predictions without radio. During the approach of the Galveston hurricane, the storm hunters knew the position of its center only when it crossed Cuba and again when it struck the Texas Coast. While it was in the Gulf, weather reports from coastal points indicated that there was a hurricane outside, moving westward, but the winds, clouds, tides, and waves at those points would have been about the same with a big storm far out over the water as with a small storm close to land. Soon after the Galveston disaster there was a growing hope that wireless messages from ships at sea would provide this vital information in time for adequate warnings.

Progress in the use of wireless at sea really was fast, although it seemed very slow to the storm hunters at the time. The first ocean-weather report to the Weather Bureau was received from the Steamship *New York,* in the western Atlantic, on December 3, 1905. It was not until August 26, 1909, that a vessel at sea reported from the inside of a hurricane. It was the Steamship *Cartago,* near the Coast of Yucatan. The master estimated the winds at one hundred miles an hour. This big storm struck the Mexican Coast on August 28, drowned fifteen hundred people and created alarming tides and very rough seas all along the Texas Coast. Thousands of people at Galveston and at many other points between there and Brownsville stood on the Gulf front and watched the tremendous waves breaking on the beaches.

Gradually the number of weather reports by radio increased and the work of the storm hunters improved. World War I and enemy submarines stopped the messages from ships temporarily, but after 1919 weather maps were extended over the oceans. Other countries co-operated in the exchange of messages and the centers of storms were spotted, even when far out of range of the nearest coast or island. Cautionary warnings were sent to vessels in the line of advance. By this means, the service of the storm hunters was of extreme value in the safety of life and property afloat as well as on shore.

By 1930 another trouble had developed serious proportions as a consequence of this efficiency in the issuance of warnings. Vessel masters soon learned that it was dangerous to be caught in the predicted path of a hurricane, and when a warning was received by radio, they steamed out of the line of peril as quickly as possible. Thus, as the storm advanced, fewer and fewer ships were in a position to make useful reports and in a day or two the hurricane was said to be "lost," that is, there were too few reports to spot the center accurately, or in some cases there were no reports at all. The storm hunters could only place it vaguely somewhere in a large ocean area. When it is impossible to track the center of a hurricane accurately, it is impossible also to issue accurate warnings.

In 1926, a hurricane crossed the Atlantic from the Cape Verde Islands to the Bahamas and threatened southern Florida. After it left the latter islands, weather reports from ships became scarce and the center was too close to the coast for safety when hurricane warnings were issued, although everybody in southern Florida knew that there was a severe storm outside. More than one hundred lives were lost in Miami and property damage reached one hundred million dollars. In 1928, another big hurricane started in the vicinity of the Cape Verdes, swept across the Atlantic, and devastated Puerto Rico and parts of southern Florida. Loss of life was placed at three hundred in Puerto Rico and at two thousand in Florida, mostly in the vicinity of Lake Okeechobee.

In these years and up to 1932, several hurricanes were "lost" in the Gulf of Mexico and citizens of the coastal areas began making demands for a storm patrol. They wanted the U. S. Coast Guard to send cutters out to search for disturbances or explore their interiors and send information by radio to the Weather Bureau. There was opposition from the forecasters—they didn't know what they would do with the cutters. If they had enough ship reports to know where to send the cutters, they would not need the latters' reports, and if they had no reports, they would not know where to send the vessels. Besides, it was the government's business to keep ships out of storms—not to send them deliberately into danger.

The season of 1933 established an all-time record of

twenty-one tropical storms in the West Indian region. Many of them reached the Gulf States or the South Atlantic Coast and the controversy about sending ships into hurricanes was resumed, resulting in legislation containing the authority, but President Roosevelt vetoed it. By 1937 the criticism of the warnings and the arguments about Coast Guard cutters began again. This time it involved Senators and Congressmen from Gulf States and finally the White House was embroiled.

In August, 1937, a delegation of citizens came to Washington and brought their complaints direct to the White House. The President arranged a conference so that the storm hunters, Coast Guard officials and others could explain again why vessels should not go out into the Gulf of Mexico to get data when the presence of a hurricane was suspected. Actually, ships were being saved by the warnings which kept them out of danger, and the criticism was based on fear of hurricanes rather than any deficiency of the warnings with respect to the coastal areas.

When the conference was held at the White House, the President was busy with other matters and James Roosevelt presided. The President had given him a note to the effect that he should receive the delegation in a most pleasant manner but that it would be dangerous and fruitless to try to send Coast Guard vessels into hurricanes.

The President's note to his son said in part:

"Make it clear that I would veto the bill again and that instead of a hurricane patrol the safest and cheapest thing would be a study of hurricanes from all of the given points on land and around the Gulf of Mexico. This might involve sending special study groups to points in Mexico, such as Tampico, Valparaiso, Tehuantepec, Yucatan, Campeche, also to the west end of Cuba and possibly to some of the smaller islands in the region. What the Congressmen and others in Texas want is study and information and it is my thought that this can be done more cheaply and much more safely on land instead of sending a ship into the middle of a hurricane."

The delegation gathered in an outer office at the White House. It happened that the Coast Guard had a new Commandant, Admiral Waesche, who had not been advised of the views of the White House, the Coast Guard, and the Weather Bureau. In the few minutes before the conference started, there was no opportunity to inform the Admiral, for he was engaged in conversation with a group of Senators and Congressmen. As soon as the conferees were assembled, James Roosevelt called on the Admiral to speak first. To the amazement of all present, he indorsed the idea in full and promised to send cutters out in the Gulf whenever a request was received from the Weather Bureau. Nobody knew what to do next, so James adjourned the conference, and after everybody had shaken hands and departed, he went back to his father to explain what had happened.

Thus began a brief period of hunting hurricanes in the Gulf of Mexico with Coast Guard cutters. During the next two seasons, the Weather Bureau forecasters notified the Coast Guard when observations were needed. In each instance a cutter left port in accordance with the agreement, but as soon as the vessel was in the open Gulf the master was in supreme command and he would not deliberately put his ship and crew in jeopardy. Cutters went out in a few cases, but most of the disturbances to be reconnoitered were crossing the southern Gulf, out of range of merchantmen on routes to Gulf ports. In sailing directly toward the center under these conditions, the Coast Guard commander would have been traveling into the most dangerous sector, and the distance he could make good in a day in rough water could not have been much larger than the normal travel of a tropical storm, certainly not a safe margin.

Irate citizens complained to Washington, first, that the Weather Bureau refused to call on the Coast Guard for observations; and, second, that the Coast Guard refused to carry out the Weather Bureau's instructions. After two or three years, no special information of any particular value was obtained and the scheme was forgotten.

In accordance with the ideas expressed by President Roosevelt, but without any support from Congress, some study groups and other special arrangements secured useful results on coasts and islands, but it was obvious after 1940 that automatic instruments for exploration of the upper atmosphere and reconnaissance by aircraft offered the best prospects for improvement in the service.

The most destructive hurricane during this period devastated large areas of Long Island and New England in September, 1938, taking six hundred lives and destroying property valued at about a third of a billion dollars. This event aroused general criticism of the storm hunters for two reasons. First, this disturbance, while it was in the West Indies and during its course as far as Hatteras, behaved like others of great intensity, but from that point northward its forward motion was without precedent. During the day when it passed into New England, its progressive motion exceeded fifty miles an hour, hence little time remained for the issue of warnings after its increased rapidity of motion was detected. Second, the people were absorbed in news of negotiations in Europe to prevent the outbreak of a world war, and storm news on the radio was largely suppressed to make way for reports of the European crisis.

Here it might be said that the storm hunters lost another battle, but it is probable that the loss of life in this hurricane would have exceeded that at Galveston in 1900 if there had been no real improvement in the warning service in the meantime.

6. THE EYE OF THE HURRICANE

>>

"—the whirlwind's heart of peace."—Tennyson

After the White House conference in 1937 about sending ships into hurricanes, some of the Weather Bureau forecasters expressed the idea that the best method of tracking hurricanes would be by airplane. What they had in mind was flying around the edge of the storm and getting three or more bearings from which the location of the center could be accurately estimated. Nothing came of the idea at the time but after World War II broke out in Europe, the talk about use of planes increased. It was the Weather Bureau's plan to contract with commercial flyers to go out and get the observations on request from the forecasters. But no one seriously considered sending planes into the centers of hurricanes. No one knew what would happen to the plane. There was no very definite information as to what the flyer would encounter in the upper layers in the region around the center.

Of course, it was known that at the surface of the earth or the sea there was a small calm region in the center—an oddity in the weather, for no other kind of large storm has such a center. The tornado may have, but it is a very small storm in comparison with a hurricane. Its writhing, twisting funnel at the vortex is hollow, according to the testimony of a few men who have looked up into it and lived to tell about it. In the tropical storm, however, nothing was known about the central winds in the upper levels. There was no proof that strong winds did not blow outward from the center up there and a plane would be thrown into the ring of powerful winds around the eye. The only way to find out was to fly into it and have a look, but there was no one at the moment who wanted to venture into it.

On the outer fringes of the hurricane, where light, gusty winds blew across deep ocean waters, stirred at the surface by giant sea swells, the hurricane hunters were fairly well satisfied with their findings. In the middle regions, where deluges of rain slanted through raging winds and low-flying clouds, the grim fact was that they knew amazingly little about what was going on in the upper layers. Their balloons sent up to explore the racing winds above were lost in thick clouds before they had risen more than a few hundred feet.

On beyond, somewhere in that last inner third of the whirlwind, the increasing gales rose to a deadly peak and torrents of rain merged with the spindrift of mountainous wave crests to blot out the view of the observer. Within this whirling ring of air and water lay the vortex. When the mariner entered, sometimes slowly, but more often suddenly, the wind and rain ceased and usually there would be no violence except the rise and fall of the sea surface, like a boiling pot on a scale which was huge in fact but small in proportion to the extent of the storm itself. The entire whirling body of air would likely be bigger than the state of Ohio; the calm central region might be the size of the city of Columbus.

Here in this inner third were the mysteries. Where could all this air go—streaming so violently around and in toward that mysterious center but never getting there? It must go up, the storm hunters argued, for what else could produce all this tremendous rainfall if not the upward rush of moist air to be cooled in the upper levels? And then, why no rain or wind in the central region? Some argued that the air must descend in the vortex, growing warmer and dry in descent, but why the descent? And finally, if the air was moving upward in all this vast area outside the calm center, what finally became of it?

Even if the storm hunters were unable to answer these questions, they could render a service of enormous value if they could track the storm and predict its movements. But they knew that the only sure way to track a hurricane over the ocean was to find its center and follow it persistently and accurately from day to day. Tests had shown that it was not practical to send ships into the storm to find its center and report by radio. Ships couldn't move fast enough. If the storm hunters had known enough about it, they might have concluded that a plane could enter the storm in the least dangerous sector and find its way swiftly to the calm center through some upper level without being hurled into the angry sea. If it reached the center of the vortex—usually called the "eye of the hurricane"—the navigator might be able to see the sky and the sun by day, the stars by night. Here the pilot might be able to figure out his position, as an ocean-going vessel does on some occasions, and that would be the location of the storm to be placed on the charts of the storm hunters in the weather office. But nobody took it seriously until after the United States got into the Second World War.

When the request for funds to hire commercial flyers in hurricane emergencies was presented to the Bureau of the

Budget, the examiners asked why the Weather Bureau didn't try to get the co-operation of the Army and Navy. Why couldn't they have their pilots carry out the flights as needed? There was some talk about it in 1942, but at that time there were no experienced Army or Navy pilots to spare.

Naturally, the military pilots who thought about flying into the eyes of hurricanes wanted to know what it was like in the upper levels and in the center. Air Force pilots who expected to go on bombing missions to Germany thought it might be more dangerous flying into the vortex of a hurricane than over an enemy stronghold with the air full of flak and Nazi fighters rising on all sides. Nobody looked upon the assignment with any enthusiasm. One discouraging fact was that the reports of shipmasters who had been in the eyes of hurricanes didn't agree very well. Few of them had the ability to describe what they saw. And those who had the ability told a story that was not reassuring. For example, one of the first was the master of the ship *Idaho,* caught in the China Sea in September, 1869, as a typhoon struck. With little of the precious sea room needed to maneuver, the ship soon was obliged to lie to and take it. Afterward, when by some miracle the ship had made its way to shore, the master calmly described his experiences while they were fresh:

"With one wild, unearthly, soul-chilling shriek the wind suddenly dropped to a calm, and those who had been in these seas before knew that we were in the terrible *vortex* of the typhoon, the dreaded center of the whirlwind. Till then the sea had been beaten down by the wind, and only boarded the vessel when she became completely unmanageable; but now the waters, relieved from all restraint, rose in their own might. Ghastly gleams of lightning revealed them piled up on every side in rough, pyramidal masses, mountain high—the revolving circle of wind, which everywhere inclosed them, causing them to boil and tumble as though they were being stirred in some mighty cauldron. The ship, no longer blown over on her side, rolled and pitched, and was tossed about like a cork. The sea rose, toppled over, and fell with crushing force upon her decks. Once she shipped immense bodies of water over both bows, both quarters, and the starboard gangway at the same moment. Her seams opened fore and aft. Both above and below, men were pitched about the decks and many of them injured. At twenty minutes before eight o'clock the vessel entered the *vortex;* at twenty minutes past nine o'clock it had passed and the hurricane returned blowing with renewed violence from the north, veering to the west. The ship was now only an unmanageable wreck."

For many years, the classic case was the obliging typhoon that moved across the Philippines with its center passing directly over the fully-equipped weather observatory in Manila. It happened on October 20, 1882. The wind which came ahead of the center was of destructive violence, reaching above 120 miles an hour in a final mad rush from the west-northwest before the calm set in. It was not an absolute calm. There were alternate gusts and lulls. The way the winds acted led the observer to think that the center was about sixteen miles in diameter. He said:

"The most striking thing about it was the sudden change in temperature and humidity. The temperature jumped from 75° to 88°. The air was saturated at 75° but the humidity dropped from 100% to 53% in the center and then rose to 100% again as the center passed. When the wind suddenly ceased at the beginning of the calm and the sun came out, many people opened their windows but they slammed them shut right away, because the hot, dry air seemed to burn the skin."

For more than fifty years after this, there were arguments

about the reasons for these changes in temperature and humidity. Some scientists claimed that they were caused merely by the heating of the sun in a clear sky and that the air which preceded and followed the center was cooled and saturated by the rain. Some of the Jesuit scientists at Manila did not agree. One weatherman showed, for example, that if they took air at 75° and 100% humidity and heated it to 88°, the humidity would fall only to about 61% and that the air at Manila at that time of year had never had such a low humidity (53%), even when the sun was shining.

The general conclusion was that the air descends in the eye of the tropical storm. At least, they were convinced that it descended in the Manila typhoon. When air descends, it is compressed, coming into lower levels where the pressure is higher. This compression causes its temperature to rise and the air then has a bigger capacity for moisture. In other words, the air becomes warmer and drier. There never has been full agreement on this question. Certainly, in some cases, the air is not warmer and drier in the center.

In later years, typhoon centers passed over other observatories and had various effects. However, one struck Formosa on September 16, 1912, and the calm center passed over the observatory long after the sun had gone down. In this case, the temperature jumped from 75° to 94° and this could not be explained by the direct heat of the sun. But there were different results in other cases and in one instance the temperature fell a little.

All of these observations were confined to ground level and what the observer could see from there or from shipboard, where he was being bounced around by violent seas and sometimes was thoroughly drenched by the mountainous waves breaking over the decks. One example was the *Idaho* in the typhoon in 1869.

A half-century later, two British destroyers were trapped

in the same region by an unheralded typhoon. Setting out for Shanghai in the early morning, they rounded the Shantung Promontory and headed across the Yellow Sea at fifteen knots, with sunlight gleaming on the water ahead. The weather looked favorable, barometer high, wind light, but it failed to stay that way very long. By ten o'clock there was a strong wind on the port beam, blowing gustily from the east, and an ominous rising sea. Reducing speed to eleven knots, the commander of the destroyer in the lead—called the *Exe*—found by dead reckoning that he was only about eight miles from land and, although he was running almost parallel to the coast, their situation was beginning to look dangerous. He had to make a decision as to his future course.

Among other disturbing factors was the design of the ships. These destroyers were of a new type, with a large forecastle which made it likely that they would drag their anchors if they tried to lie-to in a sheltered place on that exposed coast. The two ships held their course. By noon the visibility had dropped to less than a mile. The commander feared that he would be unable to identify any land he might see through the increasing gloom and concluded that his chances of finding a safe shelter among the rock-bound islands along the coast was fast becoming nil. So he signaled to the other destroyer to head fast for the open sea. In the next hour, the wind and sea mounted rapidly and he was certain that they were being overtaken by the dangerous sector of the typhoon. Now they were in real trouble!

His first lieutenant was the last of his officers out of school, so the commander asked him about the law of storms and the proper course under the circumstances. According to the latest books which the lieutenant had studied, they should have steamed toward the northwest but this would have thrown them onto a lee shore. The commander decided that there was no choice except to hold their course and run the chance of going into the dreaded center of the typhoon. So they got busy, doubly securing all movable gear and seeing that all was snug for a frightening trip into the unknown. The commander was annoyed, not so much by the battering the ship was taking as by the cheerful attitude of the lieutenant, who seemed to be looking forward to this new experience.

In this miserable situation they fought heavy gales and towering seas for hours. The other destroyer had been lost from view but now appeared close on their beam. She assumed strange attitudes in the growing darkness. "At times," the commander said, "she would be poised on the crest of a great wave, her fore part high above the sea and her keel visible up to the conning tower; the after part, also high in the roaring wind, leaving her propellers racing far out of the water. Then she would take a dive and an intervening wave would blot out this 'merry picture,' and then, to our relief as the wave passed, a mast would appear waving on the other side and then we would catch sight of her funnels and finally her hull, still above water." As darkness closed in, the crew of the *Exe* were glad they could no longer see the other destroyer for it made them vividly conscious that their own little ship was going through equally dangerous contortions.

During this time the destroyer *Exe* had suffered much damage. The upper deck had been swept clear. Much water was getting below and the pumps were choked. The commander was weary from holding on to the bridge and trying to keep his balance. The crew was frightened more than ever by the increasing power of the storm and the inexorable approach of the unknown horrors in the center.

The awful night passed in this terrifying manner, with the barometer steadily going lower, and the quartermaster straining to keep the craft on course. With powerful winds full in his face and drenched by spray, he managed to hold the ship most of the time and made the best use of her high bows. When he failed and allowed the ship to get a few points off course, the steep waves threw her on her beam ends and came crashing along the upper decks, making it a tough job to get her back with her nose against the elements, and the high bows as a sort of shield against the brutal sea. Besides, the compass light had been beaten out and in the blackness of the storm he had no way of judging the direction except by the crash of the wind and water in his face.

In a storm like this, the crew think that they are probably on their last voyage. They can feel tremendous masses of water strike with immense force and, after the shock, the vessel shivering as though the hull had given way, leaving them on the verge of diving toward the bottom of the sea. Sometimes the *Exe* was mostly out of water—they could sense it in the darkness—and then she took what they called a "belly-flopper" and every man felt sick, fearing the end had come and, after a moment, fearing just the opposite—that it would not be the end, after all, and they would have to take more of the same.

Now the lieutenant crawled out from below and, by a series of lurches between gusts, pulled himself to the side of his commander. "Things look better," he shouted. "The barometer is up a little." But soon after that he found he had made an error. He had read it an inch too high. Actually, it had dropped almost an inch in three hours, showing that the center must now be drawing near. Shortly the rain ceased and the wind dropped. At 7:00 A.M. they were passing into the vortex.

The ocean now presented a fantastic spectacle. They could see for several miles—a cauldron of steep towering cones of water with spray at the crests—a brightening sky over a chaotic sea. Some of these columns of water would clash together on different courses and produce a weird effect.

The wind became light and a few tired birds sought haven on deck. This scene lasted only ten to twenty minutes and then the dreaded squalls ahead of the opposite semicircle of the typhoon began to hit the vessel. By 7:20 A.M. the full force of the most vicious gales was bringing new miseries to the exhausted crew.

After three hours, the typhoon began to abate and the commander was feeling a little easier about his damaged ship until one of the officers reported that they had sprung a leak. The compartment containing the fore magazines was flooded and soon filled up. "So the destroyer went her way," the commander reported, "with her nose down and her tail in the air." She made it to the mouth of the Yang-tse at 11:00 A.M. Up the river a distance they found their companion destroyer. Its commander had been much impressed by the blue sky and calm in the vortex, also by the large number of birds, mostly kingfishers, that came on board.

Examination of the *Exe* showed that a part of the bottom had been battered in, shearing the rivets and opening the seams. After thinking about his good fortune in coming through the typhoon, the commander wrote in his report: "When I recall (which I can without any trouble) those awful belly-floppers the craft took, and realized by inspection in dock what amount of holding power a countersunk rivet can possibly have in a three-sixteenth of an inch plate, I wonder that I am now in this world." Actually, the commander of the *Exe* had escaped the worst of it. If he had missed the vortex and had passed through the right edge, where the forward drive of the typhoon was added to the force of the violent inner whirl, he might not have lived to tell the story. Many others have failed under similar circumstances.

Shanghai suffered severely from this typhoon. A flood in

the river and on a low-lying island drowned five thousand Chinese.

All these accounts agreed on one thing—the ring of gales around the center. Some were more violent than others but the ring was always there. On the eye of the hurricane, however, there was less agreement. A strange case was the experience on the American steamship *Wind Rush,* in October, 1930, off the west coast of Mexico. She was caught in a violent hurricane and the master suddenly saw that the ship had passed into the vortex. The second officer, in his report, said: "From 9 A.M. to 10 A.M. we were in a calm spot with no wind and smooth sea, and the sun was shining."

There have been similar instances of vessels in the vortex of hurricanes without much disturbance of the sea, but these are exceptions. Most of them have reported confused cross seas, described as "pyramidal" or "tumultuous." In November, 1932, the master of the British steamship *Phemius,* on a voyage from Savannah to the Panama Canal, was so unfortunate as to become entangled in the outer circulation of a late-season hurricane moving westward in the Caribbean Sea. It turned sharply northward and the *Phemius* was trapped by the ring of fierce gales in the central region. She rolled through an arc of 70° while the gusts came with such force that the funnel was blown away. The master put the wind at two hundred miles an hour. Hatches were blown overboard like matchwood, derricks and lifeboats were wrecked, and the upper and lower bridges were blown in. The ship was rendered helpless and was carried with the hurricane in an unmanageable condition.

Twice the *Phemius* drifted into the vortex, with high, confused seas and light winds. The second time the vessel was besieged by hundreds of birds. They took refuge in every part of the ship but lived only a few hours. Driving toward the coast of Cuba, the hurricane ravaged the town of Santa

Cruz del Sur, hurling a tremendous storm wave across all the low ground, engulfing the town, and drowning twenty-five hundred persons out of a population of four thousand. The *Phemius* was left behind in a helpless condition and was taken in tow by a salvage steamer.

The width of the eye of a hurricane commonly varies from a few miles to twenty or twenty-five. The smallest known was entered by a fishing boat, the *Sea Gull*, in the Gulf of Mexico, on July 27, 1936. The master, Leon Davis, was fishing a few miles east of Arkansas Pass, Texas, when he became involved in a small hurricane. "Suddenly," Captain Davis said, "the wind died down, the sun shone brightly and the rain ceased. For a space of about a mile and a half, a clear circular area prevailed; the dense curtain of rain was seen all around the edge of the circle; and the roar of the wind was heard in the distance." On the other hand, one of the largest eyes yet known attended a big hurricane in October, 1944. It blasted its way across Cuba and entered Florida on the west coast, near Tampa. As it neared Jacksonville, the calm center was stretched out to the remarkable distance of about seventy miles. This was a kind of freak; some of the storm hunters thought that it had been distorted and finally drawn into an elongated area by its passage over the western end of Cuba.

All of the available records of this kind were consulted in due time by the men who were assigned to the perilous duty of flying military planes into the vortices of hurricanes in the West Indies and into typhoon centers in the Pacific. But one of the best of these reports—of weather and sea conditions observed on many ships caught at the same time in the central region of a big typhoon—was not available until long after it happened. The Japanese kept it secret for seventeen years.

The reason for keeping the data secret was the fact that

while on grand maneuvers, the RED Imperial Japanese Fleet was outmaneuvered by a pair of typhoons and was caught in the center of one of them and severely damaged. It happened in 1935 and was not reported for publication in America until 1952.

Just how this happened is not altogether clear. It was in the middle of September, 1935, when the first typhoon appeared, northwest of the island of Saipan. It increased in fury as it moved slowly toward Japan. On the twenty-fifth it crossed western Honshu and roared into the Sea of Japan, headed northeastward in the direction of the Japanese Fleet. Soon after this, it dissipated. Before it weakened, however, another typhoon had formed near Saipan and started toward Japan. It turned more to the northward than the first typhoon and missed Japan altogether. As it approached Honshu, late on the twenty-fifth, the RED Imperial Fleet was passing through the Strait of Tsugaru into the Pacific—squarely in front of the typhoon center.

The logical explanation for this apparent blunder is that the commanders wanted more sea room than was at hand in the northeast Sea of Japan to maneuver in the first typhoon and hoped to get well out in the open Pacific before they could be cornered by the second one. But it turned northeastward and went faster and farther out in the Pacific than they had expected. In fact, its forward motion was more than forty miles an hour in these last hours before its furious winds surrounded the fleet.

It was a bad calculation for the naval commanders and perhaps for the weather forecasters. Among the latter, H. Arakawa, one of the foremost typhoon students in Japan, was then on the staff of forecasters in the Central Meteorological Observatory in Tokyo. He was in part responsible for the predictions. In 1952 he made the report which was published by the U. S. Weather Bureau early in 1953. Taking the view of the weatherman, Arakawa said that although the damage to the fleet was unfortunate, there was *fortunately* a magnificent collection of reports from the central region of the typhoon for scientific study.

The fleet was caught in the central part of the big storm on the twenty-sixth of September. Among the ships involved, many of them damaged, were destroyers, cruisers, aircraft carriers, a seaplane carrier, mine-layers, transport ships, submarines, torpedo boats, and a submarine depot ship. The fleet suffered damage mostly from the tremendous waves in the right rear quadrant of the typhoon. Here the rapid forward motion of the storm was added to the wind circulation and the seas were driven to excessive heights. In his report, Arakawa had a footnote: "The bows of two destroyers, *Hatsuyuki* and *Yugiri*, were broken off as a result of excessive storm waves, and many officers and sailors were lost."

In the calm center, the clouds broke and faint sunlight came through. The diameter of the eye was nine or ten miles. To the right of the eye, some of the waves measured more than sixty feet in height. The maximum roll of the ships in this area—the total angle from port to starboard—reached 75° on some of the ships. The wind was steadily above eighty miles an hour; the gusts were not measured but probably went as high as 125 miles an hour.

Many of the ships took frequent observations while in the typhoon and the data would have been extremely valuable if released to the storm hunters at that time, but when the report was published in 1953 a great deal of new data had been obtained by airplane, both at the surface—where Arakawa's observations were confined—and at higher levels. It was a little more than nine years after this Japanese incident when the U. S. Third Fleet was caught in a typhoon east of the Philippines and suffered at least as much damage as the Japanese in 1935.

One fact is clear. For many years the storm hunters had been gathering information about hurricane and typhoon centers from observations on land and sea but they knew very little of what went on there in the upper air. World War II brought a new era.

7. FIRST FLIGHT INTO THE VORTEX!

>>>

"Whirlwinds are most violent near their centers."—Euripides

After war broke out in Europe in 1939, the job of finding and predicting hurricanes became steadily more difficult. Ships of countries at war ceased to report weather by radio and fewer vessels of neutral nations dared to risk submarine attack. After Pearl Harbor, the American merchant marine also stopped their weather messages and the oceans were blanked out on the weather maps. Already the British had been confronted by the lack of weather reports from the Atlantic and the seas around the British Isles, and this was extremely serious in their fight against Nazi air power.

Notwithstanding the alarming scarcity of planes for military purposes, the British were forced to send aircraft on routine weather missions. They usually flew a track in the shape of a triangle—for example, one leg of the triangle northwestward until well out at sea, a second leg southward across the ocean about an equal distance, and the last leg back to home base. Other triangles were flown over Europe and back and over the North Sea. As time went on, the pilots of these observation planes gained much experience in flying the weather, including some fairly bad storms, but no one had occasion to fly into a hurricane. There was a good deal of talk about the situation in the United States in 1942, however, because of the danger that the West Indian region might become a theater of war, if the Nazi armies gained control of West Africa and attacked the United States by air, across Brazil and the Caribbean.

With this threat from the southeast, the United States took action, which was a repetition of the events during the Spanish War in 1898. Military weather stations were set up in the West Indies and aircraft were prepared to fly weather missions in the area. At the same time, the United States was getting ready to ferry planes across the South Atlantic via the Caribbean, the South American Coast and Ascension Island. It was very definitely evident early in 1942 that hurricanes might play a critical role if the West Indies became a theater of war. By 1943, however, there were two surprising turns of affairs. The Allies invaded Africa late in 1942 and the first flight into a hurricane center, unscheduled and unauthorized, came in 1943.

The first to fly into the vortex of a hurricane was Joseph B. Duckworth, a veteran pilot of the scheduled airlines, who was at the time a colonel in the Army Air Corps Reserve, in command of the Instrument Flying Instructors School at Bryan, Texas. It was one of those rare combinations of circumstances by which the man with the necessary skill, experience, daring, and inclination happened to be at the right place at the right time. With a full appreciation of the danger, he flew a single engine airplane deliberately into the hurricane and proceeded on a direct heading into the calm center, looked around, and flew back to Bryan. Spotting his weather officer, he bundled him into the back seat and duplicated the feat immediately!

Joe Duckworth was born in Savannah, Georgia, on September 8, 1902, which, incidentally, was the anniversary date of the terrible Galveston disaster of 1900, and it was a hurricane at Galveston into which he flew in 1943. Joe's mother was Mary Haines, a Savannah girl. His father, Hubert Duckworth, was a naturalized Englishman who had been sent to the States to take over the American cotton offices of Joe's grandfather, after whom he was named. When Joe was two years of age, the family moved to Macon, Georgia, where his father was vice-president of the Bibb Manufacturing Company.

Joe's first memory of anything connected with aviation was when his parents took him to the fair grounds at Macon to see Eugene Ely fly in an early Wright-type biplane. The wind was not right for a flight. Pilots were cautious in those days and Ely didn't go up. Joe and his parents were looking at the plane when his father remarked, "You know, some day they will be carrying passengers in these things." His mother answered, "Don't be silly, Hubert, you might as well try to fly to the moon." Joe had a vague idea at the time that he would like to fly when he grew up. Long afterward, he did. He says, "Many times in the nineteen thirties I captained an Eastern Airlines plane over Macon and looked down on the old fair grounds and recalled the thrill I had on seeing my first airplane and the remarks of my mother and father."

After his father died in 1914, Joe attended Woodberry Forest School in Orange, Virginia, for three years and then went for two years to Culver Military Academy in Culver, Indiana, graduating from there in 1920. In the meantime, his mother had moved to Atlanta and he continued his education for two years at Georgia Tech. and one year at Oglethorpe University. Nothing he did would take flying out of his mind and he finally gained admission to the Flying Cadets. After going through both Brooks and Kelly Fields as Cadet Captain, he was graduated in 1928, the happiest year of his life. Later, while flying for Eastern Airlines, he got a law degree from the University of Miami.

With basic training of the kind that young Duckworth received as a Cadet, he was not fitted to fly into a hurricane or into any sort of really bad weather. Military operations at that time were strictly visual or "contact." The problem was not how to get through bad weather—thunderstorms, low overcast, fog, for example—but how to keep out of it. There were few flight instruments, and there was no instrument flying training. At that time, dirigibles were thought by many leaders in aeronautics to have the best passenger-carrying possibilities for the future. Steel had just replaced wood in fuselages and airplanes in general had earned the description "heavier-than air." On the other hand, the world had been electrified by Lindbergh's flight to Paris in 1927 and other "stunt" flights became numerous. Another thrilling piece of news was Admiral Byrd's flight to the South Pole in 1929.

Trial freight-carrying runs were being flown by the Ford Motor Company from Detroit to Chicago and from Detroit to Buffalo, and Joe heard that a young man could get tri-motor flight time as a co-pilot two days a week, provided he worked four days in the factory. Duckworth headed for Detroit. After getting on the job with Ford, he had his first serious run-in with clear ice, or freezing rain. The plane barely made South Bend Airport, coming in at high speed with a load of ice on the wings. Fifteen years later, the pilot on instruments would have climbed quickly into the warmer air at higher levels and then worked his way down to destination, but instrument flying was unknown at the time.

In the spring of 1929, Joe went with the Curtis Wright

Flying Service as their first instructor, at Grosse Ile, near Detroit. They were starting out to set up a nation-wide chain of bases with the idea of teaching everyone to fly. The plan was successful at first and in the fall Joe opened a branch at Atlanta, just as the stock market broke wide open. The slump in business that followed in 1930 caused general failure in the flying services. In December, Joe saw that the Atlanta branch was going out of business, and he went to work as a pilot for Eastern Air Transport, now Eastern Airlines, and remained with the company for ten years. At first he flew mail planes with parachutes but no passengers.

Even then there was no such thing as flying the weather. On his first mail flight, he got some pointed advice from the operations manager. He told Joe to be "sure to be on the look out for a reflection of the revolving radio beacon on the cloud ceiling and the moment you see such an apparition, you must get down immediately in an emergency field. If you let the overcast close down on you, you are strictly out of luck." Airplanes were a long way then from being equipped to fly into hurricanes.

What little was known at that time about the temperature, pressure, and humidity in the upper air was secured by kites sent up daily at a few places. They were box kites, carrying recording instruments and flown by steel piano wire. Observers let them rise and pulled them in by reels and, after examining the records, sent the data to the weather forecasters. This was a slow process and, besides, it was becoming dangerous around airports where the data were needed most. A long piano wire in the sky was a serious hazard for aircraft. After 1931 this method was abandoned, and pilots under contract to the Weather Bureau attached weather-recording instruments to their planes and ascended to a height of three miles or a little higher, and on return gave the records to the weather observer, who worked them up and wired the results to the forecasters. Army and Navy pilots carried out similar missions at military bases. This plan worked fairly well. The flights were made early in the morning but when the weather was bad and the data would have been most useful, the planes were obliged to remain on the ground.

Gradually, beginning about 1932, airline pilots began more and more intentional flights "on instruments," that is, operating in clouds without visual reference to ground or horizon. Reliability of schedules was an economic necessity. Navigation by radio was becoming more of a commonplace and, by experiment and self-teaching, by 1940 airlines were flying almost all kinds of en route weather, including thunderstorms.

In 1940, Joe's thoughts turned to the Army Air Corps, in which he held a reserve commission as Major. It looked as though war might come to the United States, so in November of that year he resigned from Eastern to enter active duty—probably the first airline pilot to do so. Assigned to the Training Command, he never got overseas—but what he did in teaching instrument flying throughout the Air Corps is still acknowledged and appreciated by thousands of wartime pilots. He received literally hundreds of letters expressing their gratitude, some of them declaring that the training they had received had literally saved their lives on many occasions.

Joe found a serious lack of instrument flight training in the Air Corps, due to the frenzied expansion of training for war. And, as Joe said, "You couldn't call off the war when the weather was bad!" He set out to make his wartime mission the remedying of this situation, and the record will show he did a monumental job. Cutting "red tape" wherever possible, experimenting, lecturing, and writing a whole new system of instrument flying training, he and his chosen assistants culminated two years of intensive effort by establishing an instrument flying instructors school at Bryan, Texas, in February of 1943. During the next two years, the school provided over ten thousand highly qualified instructors to the Army Air Forces, and attained a solid reputation which is not forgotten today. Joe's instructors flew all types of weather—anywhere—and at the same time piled up a safety record unheard of at the time. The manuals they developed are still, in principle, the standard of today's Air Force.

Joe's school taught, through novel and thorough techniques, two things. First, that there is no weather, except practically zero-zero landing conditions, that cannot be flown by the competent instrument pilot, with proper equipment. Second, that the safety *and* utility of both military and commercial flight depend almost wholly on the competence of the pilot in instrument flying.

Thus it came about that the first flight into a hurricane center was not the result of a sudden notion but of years of intensive training in flying the weather, including storms, and the flyer who did it was probably the most expert in the world at getting safely through all kinds of weather. Looking at it from this point of view, it is not strange that there was a rather amusing sequel to this story, involving the other instructors at Bryan, Texas. But first we come to the story of the history-making flight by Colonel Duckworth.

Early on the morning of July 27, 1943, Joe came out to have breakfast at the field. The sky at Bryan was absolutely clear and it did not seem to promise any kind of weather that would try the mettle of men whose business it was to fly in stormy conditions. Someone at the table said he had seen a report that a hurricane was approaching Galveston. Joe was immediately attentive. Sitting opposite him was a young and enthusiastic navigator, the only one at the field, Lieutenant Ralph O'Hair.

Thinking again about the fact that no one had flown a hurricane and that it ought to be easy because of the circular flow, Joe suggested to Ralph: "Let's go down and get an AT-6 and penetrate the center, just for fun." He said it would be "for fun" because he felt sure that higher headquarters probably would not approve the risk of the aircraft and highly trained personnel for an official flight. There were three or four newly arrived B-25's at the field but Duckworth had not had the time to check out in one of them and therefore could not fly a B-25 (a twin-engine airplane) without going through some formalities. Use of the AT-6, of course, involved the danger that its one engine might quit inside the hurricane and they would be in trouble.

Lieutenant O'Hair was quite willing—enthusiastic in fact—and the pair gathered such information as was available about the hurricane and made ready for the flight. They took off in the AT-6 shortly after noon. The data on the storm had been rather meagre. Two days before, Forecaster W. R. Stevens at New Orleans had deduced from the charts that a tropical storm was forming in the Gulf to the southward. He drew his conclusion almost solely from upper air data at coastal stations, for no ships were reporting from the Gulf. On the twenty-sixth, Stevens had correctly tracked it westward toward Galveston (quite a feat in view of the lack of observations) and warnings had been issued in advance.

On the morning of the twenty-seventh, this small but intense hurricane was moving inland on the Texas Coast, a short distance north of Galveston, and by early afternoon the winds were blowing eighty to one hundred miles an hour on Galveston Bay and in Chambers County, to the eastward of the Bay. Houston and Galveston were in the western or less dangerous semicircle, a favorable condition for the flight from Bryan to Houston. Soon after leaving Bryan, the venturesome airmen were in the clouds on the outer rim of the storm—with scud and choppy air—and shortly after they ran into rain. Precipitation static began to give them trouble in communications but there was no other serious difficulty.

As they approached Houston, the air smoothed out, the static leaked off the plane, the radio was quiet, and the overcast grew darker. They called Houston. The airways radio operator was surprised when they said their destination was Galveston.

"Do you know there is a hurricane at Galveston?" the operator asked.

"Yes, we do," said O'Hair. "We intend to fly into the thing."

"Well, please report back every little while," the operator requested. "Let me know what happens." Evidently, he wanted to be able to say what became of the plane if they went down in the storm.

At this point Joe's mind began to run back over some of the lectures the flight instructor had given and recall how they had stressed the fact that a pilot should always have an "out," even if it meant taking to a parachute. He wondered what it would be like to use a parachute in a hurricane. They were flying at a height of four thousand to nine thousand feet.

As they approached the center, the air became choppier again and he said afterward that they were "being tossed about like a stick in a dog's mouth," without much chance of getting away from the grip of the storm. Checking on the radio ranges at Houston and Galveston, they flew over the latter and then turned northward. Suddenly, they broke out of the dark overcast and rain and entered brighter clouds. Almost immediately, they could see high walls of white cumulus all around the circular area in the center and, below them, the ground and above the sky quite clearly. The plane was in the calm center. The ground below was not surely identified but it seemed to be open country, somewhere between Galveston and Houston. They descended in an effort to get their position more clearly but the air became rougher as their altitude decreased. This led Duckworth to the conclusion that the eye of the hurricane was like a "leaning cone," the lower part probably being restricted and retarded by the frictional drag of the land over which the storm was passing. They flew around in the center a while and then took a compass course for Bryan.

Once out of the center, the plane went through, in reverse, the conditions the fliers had experienced on the way in, arriving at the air field at Bryan in clear weather. When they got out of the plane, the weather officer, Lieutenant William Jones-Burdick, came up and said he was very disappointed that he had not made this important flight.

Duckworth said, "OK, hop in and we'll go back through and have another look." So he and the weather officer flew into the calm center again and looked around a while. The weather officer kept a log from which the following excerpts are taken, beginning with their entry into dense clouds on the way into the hurricane. The time given here is twenty-four-hour clock. Subtract 1200 to get time (P.M) by Central Standard.

1715 Heavy rain, strong rain static.
1716 Rain continues but static only moderate. Some crash static intermittently.
1720 Getting darker, cloud more dense, rain very heavy, turbulence light. Rain static building up, blocking out Galveston radio range intermittently.
1725 Turbulence light to moderate, rain very heavy.
1728 Altitude 7300′. Free air temperature 46°, cloud getting somewhat lighter.

1730 Rain less heavy, cloud much lighter, ground visible through breaks. Surface wind apparently South Southeast.

1735 Crossed east leg of Galveston range and changed course to 330°.

1740 Now flying in thick cloud. Turbulence smooth to light.

1743 Turbulence moderate.

1744 Turbulence moderate to severe.

1745 Sighted clear space ahead and to the left.

1746 Now flying in "eye" of storm. Ground clearly visible, sun shining through upper clouds to the west. Circling to establish position. Surface wind South.

1753 Still circling. Altitude 5000′, temperature 73°.

1800 Headed west for Houston. Cloud very dense, rain light, turbulence moderate, intermittent precipitation static.

1805 Apparently in a thunderstorm. Altitude 5500′. Heavy rain, turbulence moderate to severe. Free air temperature now 46°.

1815 Changed heading to 10°. Rain light to moderate. Turbulence light.

1825 Headed 330°. Rain very light, turbulence almost smooth. Apparently flying between thick cloud layers.

1835 Altitude 5500′. Broken stratocumulus clouds below, high overcast of altostratus above.

1836 Breaking out into the open with high altostratus deck above.

1900 Landed at Bryan. Sky clear to the northwest.

One sequel to this story was Duckworth's discovery, a year later, that after these flights into the center, some of his instructors and supervisors who were checked out in B-25's had sneaked out and flown the same hurricane! They were afraid to tell him about it at the time, for they did not have permission to do it, but he accidently learned about it the next year, when he overheard some of them talking about their trips into the storm.

Altogether, Joe did not consider his flights into the hurricane to be as dangerous as some of his other weather flights. Only two things worried him at the time, the heavy precipitation static and the possibility that heavy rain might cause the engine to quit. Afterward, when pilots began to fly hurricanes as regular missions, the effect of torrential rain in lowering engine temperatures proved to be a real hazard and they had to take special precautions on this account.

Considering his hurricane penetration a routine weather flight at the time, Joe thought nothing more about it until he read a story in a Sunday paper, several weeks later. Then he had a telephone call from Brigadier General Luke Smith, at Randolph Field, who asked him to come down, and surprised him by saying that he knew of the incident. At Randolph, the General said that Joe was being recommended for the Distinguished Flying Cross. This never went through but later Joe did receive the Air Medal.

There were several amazing features about these flights into the vortex. First, they justified Duckworth's unswerving confidence in his ability to fly safely through a hurricane; second, at the level of high flights there was a remarkable absence of violent up drafts or turbulence; third, they showed that quiet air in the center extended at least to heights of a mile to a mile and a half, and that at those levels the air in the center was much warmer than the air in the surrounding region of cloud, rain, and high winds. Joe is sorry now he did not organize his flight to get better scientific data. He believes his air temperature gauge probably was inaccurate. But, as he says, "It was just a lark—I didn't think anybody would ever care or know about it!"

This demonstration was followed by an increasing number

of penetrations by aircraft into the eyes of tropical storms, not all of which, by any means, were as uneventful as the flights by Duckworth and his fellow officers. After years of experience, the military services involved in flying hurricanes developed a technique which was essentially the same as that used by Duckworth in this first flight; that is, penetrate into the western semicircle and then into the center or eye from the southwest quadrant.

8. THE HAMMER AND THE HIGHWAY

Bellowing, there groan'd a noise
As of a sea in tempest torn
By warring winds. The stormy blast of Hades
With restless fury drives the spirits on.

—Dante

During the first half of the present century there was a tremendous growth in population, industry, truck-farming, citrus-growing, boating, and aviation on the Gulf and South Atlantic coasts of the United States. This brought new worries to the hurricane hunters and forecasters.

By the beginning of the century, most of the older cities and port towns in this region had been hit repeatedly by tropical blasts. Insecure buildings had been eliminated. From bitter experience, the natives knew what to do when a storm threatened. They had built houses and other structures to withstand hurricane winds, placing nearly all of them above the highest storm tides within their memories. Down in the hurricane belt of Texas and Louisiana, a sixtypenny nail was known as a "Burrwood finishing nail." The town of Burrwood, at the water's edge on the southern tip of Louisiana, had no frame buildings that had survived its ravaging winds and overwhelming tides except those which were put together with spikes driven through heavy timbers.

Learning to deal with hurricanes takes a lot of time. Most places on these coasts have a really bad tropical storm about once in ten or twenty years. And so it happened that while the population was increasing rapidly in the years from 1920 to 1940, many thousands of flimsy buildings were constructed in the intervals between hurricanes. Too many were built near the sea, where they would be wrecked by the first big storm wave. To build near the water is tempting in a hot climate. And so it happened that after 1920, widespread destruction of property and great loss of life attended the first violent blow in many of these rapidly growing communities.

Newcomers—and there were many—didn't know what to do to protect life and property. After the first calamity, they were alarmed by the winds which came with every local thundershower and they were likely to flee inland in great numbers whenever there was a rumor of a hurricane. Here they became refugees, to be fed and sheltered by the Red Cross and local welfare organizations. By the middle thirties, this had become a heavy burden on all concerned. To get things under control, local chapters of the Red Cross were formed and other civic leaders joined in seeing that precautions were taken when required, and panics were averted at times when no storm was known to exist. But when warnings were issued by the Weather Bureau, coastal towns were almost deserted. The greatest organized mass exodus from shore areas in advance of a tropical storm occurred in Texas, in 1942. On August 30, a big hurricane with a tremendous storm wave struck the coast between Corpus Christi and

Galveston. It had been tracked across the Caribbean and Gulf, and ample warnings had been issued. More than fifty thousand persons were systematically evacuated from the threatened region and though every house was damaged in many towns, only eight lives were lost.

All of this brought heavy pressure on the hurricane hunters and forecasters to be more accurate in the warnings, to "pinpoint" the area to be seriously affected, and to defer the hoist of the black-centered red hurricane flags until those responsible were reasonably sure of the path the storm would take across the coast line. Thus, the warnings actually became more precise, but in some instances the time available for protective action was correspondingly reduced.

Precautionary measures must be carefully planned. The force of the wind on a surface placed squarely across the flow of air increases roughly with the square of the wind speed. For this reason, it is a good approximation to say that an eighty-mile wind is four times as destructive as a forty-mile wind. A 120-mile wind is nine times as destructive. In order to lessen property damage, residents of Florida and other states in the hurricane belt prepared wooden frames which could be quickly nailed over windows and other glassed openings. These devices proved to be very effective. In some cases it was a dramatic fact that, if two houses were located side by side, the one with protective covers on windows and other openings escaped serious damage while the other house soon lost a window pane and then the roof went off as powerful gusts built up strong pressures within the building. At the same time that this protection was applied on the windward side, openings on the leeward side (away from the wind) helped to reduce any pressures that built up in the interior.

As these experiences became common after 1930, wood and metal awnings were manufactured so that they could be lowered quickly into position to protect windows of residences. Business houses stocked wooden frames that could be fastened in place quickly to prevent wholesale damage to plate glass windows.

Many other measures were taken hastily when the emergency warnings were sent out. One, for example, was a check by home owners to make sure that they had tools and timbers ready to brace doors and windows from the inside if they began to give way under the terrific force of hurricane gusts. They had learned that with a wind averaging eighty miles an hour, say, the gusts are likely to go as high as 120 miles an hour and it is in these brief violent blasts, so characteristic of the hurricane, that the major part of the wind damage occurs.

In addition, the experienced citizen prepares for hours when water, lights, and electric refrigeration will fail. He knows, too, that these storms have a central region, or eye, where it is calm or nearly so, and he does not make the often-fatal mistake of assuming that the storm is over when the calm suddenly succeeds the roaring gales. He wisely remains indoors and closes the openings on the other side of the house, for the first great gusts will come from a direction nearly opposite that of the most violent winds which preceded the center.

In the early thirties, the hurricane forecasts for the entire susceptible region were still being made in Washington, having been begun there in 1873. Weather reports were coming in season from observers at land stations in the West Indies, mostly by cable. From many places the cable messages went to Washington via Halifax. Ship's weather messages came by radio to coastal stations on the Atlantic and Gulf, and from there to Washington by telegraph. Twice a day these reports were put on maps and isobars, and pressure centers (highs and lows) were drawn.

In general, the same system is used today. Arrows show the direction and force of the wind at each of many points; also the barometer reading, temperature, cloud data, and other facts are entered. Conditions in the upper air are shown at a few places where balloon soundings are made. As the map takes shape, it begins to show the vast sweep of the elements across the southern United States, Mexico, Central America, and all the region in and around the Caribbean Sea and the Atlantic. In these southern regions, the trade winds, coming from the northeast and turning westward across the islands and the Caribbean, bring good weather to the edge of the belt of doldrums.

This is the lazy climate of the tropics, in the vast spaces where the bulge of the earth near the Equator seems to give things the appearance of a view through a magnifying glass. In the distant scene, islands are set off by glistening clouds hanging from mountain tops. White towers of thundery clouds push upward here and there over the sea, in startling contrast to the blue of the sky and water. Nature seems to be at peace but the trained weather observer may see and measure things that are disturbing to the weather forecasters when put together on a weather map of regions extending far beyond any single observer's horizon.

Here and there in this atmosphere that seems so peaceful an eddy forms and drifts westward in the grand sweep of the upper air across these southern latitudes. These temporary swirls in the atmosphere, some of which are called "easterly waves," are marked by a wave-like form, drifting from the east. The wind turns a little, the barometer falls slightly, the clouds increase temporarily, but nothing serious happens and the eddy passes as better weather resumes. This goes on day after day and week after week, but during the hurricane season the storm hunters are always on the alert.

All this work of charting the weather day by day and week by week is not wasted if no hurricane develops. Planes take off every day from southern and eastern airports, carrying passengers to Bermuda, Nassau, Trinidad, Cuba, Jamaica, Mexico, and Central and South America. The crews stop at the weather office to pick up reports of wind and weather for their routes and at destination. The weather over these vast expanses of water surface is reported and predicted also for ships at sea. And when a storm begins to develop, ships and planes are among the first to be notified.

Sooner or later, one of these swirling waves shows a definite center of low pressure, with winds blowing counterclockwise around it. Now the modern drama of the hurricane begins. In the region where these ominous winds are charted, radio messages from headquarters ask for reports from ships—every hour, if possible—and weather offices on islands are asked to make special balloon soundings of the upper air and send reports at frequent intervals. Warnings go out to vessels in the path of the storm as it picks up force. Alert storm hunters in Cuba and other countries are contacted to discuss the prospects, to furnish more frequent reports, and to assist in warning the populations on the islands.

On the coast of the United States, excitement is in the air. Conversations in the street, offices, stores, homes, everywhere, turn to the incipient hurricane, and become more insistent as the big winds draw nearer. And finally the hour comes when precautions are necessary. By this time, business in the threatened area is at a standstill. The situation is like that during world-series baseball games and almost as dramatic as that which follows a declaration of war. Few people have their minds on business. At this point, the reports of storm hunters and the decisions of forecasters involve the immediate plans of hundreds of thousands of people, large costs for protection of property, and the safety of human life along shore and in small craft on the water.

Some of the men and women who came down to the weather and radio offices this morning know now that they will not go home tonight. There will be an increasing volume of weather reports, the rattle of teletypewriters will become more insistent, the radio receivers will be guarded by alert men growing weary toward morning, planes will be evacuated from airports in the threatened region and flown back into the interior, and the businessman will go home early and get out the frames he uses to board up the windows when a hurricane is predicted. The Navy may take battleships and cruisers out of a threatened harbor, so that their officers will have room to maneuver.

Under these dramatic conditions the hurricane comes toward land with good weather in advance—sunny by day and clear at night. The native fears the telltale booming of the surf and feels concerned about the fitful northeast breezes. In time there are lofty, thin clouds, spreading across the sky in wisps or "mares' tails" of cirrus—composed of ice crystals in the high cold atmosphere far above the heated surface of the subtropics. A thin veil forms over the sky. At the end of the day, red rays of the lowering sun cast a weird crimson color into the cloud veil, reflecting a scarlet hue over the landscape and the sea. For a few minutes the earth seems to be on fire. To the visitor, it is a beautiful sunset. To the native, it is alarming, and in some parts of the Caribbean it is terrifying as an omen of the displeasure of the storm gods. In these dramatic situations the head forecaster makes his decision.

Also, during these nervous hours, representatives of the Red Cross begin arriving on the scene. At the same time, crews assigned to duties of repairing telephone, telegraph, and power lines are sent to the threatened area by their respective companies. As soon as the storm has passed, these men will be ready to go to work.

At this juncture it is probable that strange things will happen. Against the stream of refugees moving away from the coast, there are always a few adventurers who come from more distant places to see the full fury of a great hurricane roaring inland from the sea. At first they thrill to the crash of tremendous waves breaking on the coast and hurling spray high into the screaming winds. But when the rain comes in torrents, striking with the force of pebbles, and beach structures begin to collapse and give up their components to wind and sea, the curious spectator has had enough. Hurriedly he seeks refuge and begins to wonder fearfully if it will get worse. It does. He soon realizes that what he has seen is only the beginning.

As the full force of the blast strikes a coastal city, the scene goes beyond the power of words to describe. Darkness envelops everything, with thick, low-flying clouds and heavy rain acting like a dense fog to cut down on visibility. The air fills with debris, and with the roar of the winds and the crash of falling buildings. Power lines go down, and until the current can be cut off, electric flashes throw a weird, diffuse light on the growing chaos. In the lulls, the shrieks of fire apparatus and ambulances are heard until the streets become impassable.

Most of these great storms move forward rather slowly—often only ten to twelve miles an hour. A boy on a bicycle could keep ahead of the whirling gales if the road took him in the right direction. Automobiles carrying news reporters and curious people travel the highways far enough in advance to avoid falling debris, listening to the radio broadcasts from the weather office to learn of the progress of the storm. Of all places, the most dangerous are on the immediate coast and on islands near the coast, where the combination of wind and wave is almost irresistible. But even here an occasional citizen chooses to remain, in spite of the warnings, and when he finally decides to leave it may be too late to get out and no one can reach him. There have been many instances of men being carried to sea, clinging to floating objects, and after describing a wide arc under the driving force of the rotary winds, being thrown ashore miles away from home. But in other cases, people are trapped and drowned in the rising waters. In 1919, at Corpus Christi, warnings were issued while many residents were at their noon meal, on a Sunday. Many delayed to finish eating while the only road to higher ground was being rapidly flooded. Of these 175 were drowned.

The native knows all of the preliminary signs well enough, and it is not necessary to urge him to take precautions after the moment when the ominous gusts of the first winds of the storm are felt. He has been in these situations before and has looked out to see palm trees bent far over and the rain beginning to blot out the view as the fingers of the gale seemed to begin searching his walls and roof for a weak spot. Many prefer not to stay and watch. They board up their windows and doors and go back to a safer place in the interior. And so this is the time when the sound of the hammer is heard and streams of refugees are seen on the highways.

In the early thirties, the increasing population in the hurricane states caused an annoying shortage of communications in storm emergencies. For many years the Washington forecasters had sent warnings by telegraph and the men in weather offices along Southern coasts had talked to each other by telephone, to exchange notes and opinions, but there were frequent delays and failures after 1930 because, when a hurricane approached the coast, the lines became congested with telephone calls and telegraph messages between relatives and friends worrying about the dangers, and by residents making arrangements for evacuation, in addition to emergency calls of many other kinds.

In 1935, the Weather Bureau found a very good answer to the communication shortage in emergencies. A teletypewriter line called the "hurricane circuit," running around the Gulf Coast and Atlantic Coast of Florida, was leased on July 1, with machines in all weather offices. Another line was installed between Miami and Washington and eventually extended to New York and Boston. No matter how congested the public lines became, the weather offices were able to exchange messages and reports without any delay. At the same time, three hurricane forecast offices were established in the region—at Jacksonville, New Orleans, and San Juan. After that time, the Washington office issued forecasts and warnings of hurricanes only when they came northward to about 35° north latitude and from there to Block Island, where the Boston office took over.

The first violent tropical storm to strike the coast of the United States after the hurricane circuit was set up came across the Florida Keys on Labor Day, 1935. It was spotted in ship reports and by observations from Turks Island on August 31 as a small storm. It moved westward not far from the north coast of Cuba on September 1 and turned to the northwest on September 2, having developed tremendous violence.

This hurricane is worth noting, for its central pressure, 26.35 inches, was the lowest ever recorded in a tropical storm at sea level on land anywhere in the world. The average pressure at sea level is about 29.90 inches. The biggest tropical storms have central pressures below 28.00 inches, but very rarely as low as 27.00 inches.

The strongest winds around the center of the Labor Day hurricane probably exceeded two hundred miles an hour. About seven hundred veterans of World War I were in relief camps at the point where the center struck. A train was sent from Miami to the Keys to evacuate the veterans ahead of the storm, but it was delayed and was wrecked and thrown off the tracks as the veterans were being put aboard. The loss of life among veterans and natives on the Keys in the immediate area was nearly four hundred. There was much criticism in the press. In 1936, a committee in Congress carried on a long investigation of the circumstances which led to the establishment of the relief camps in such a vulnerable position, the failure of the camp authorities to act on warnings from the Weather Bureau, and the delay of the rescue train. There was much talk in the committee of increasing the Weather Bureau's appropriations, to enable it to give earlier warnings, but nothing came of it.

The new teletypewriter circuit served well. After this violent hurricane crossed the Keys, it went through the eastern Gulf and then passed over Western Florida and overland to Norfolk. In spite of intense public excitement, communications between weather offices were maintained without serious interruption. This improved service continued in the years that followed. Radio circuits to the West Indies and a teletypewriter circuit to Cuba by cable helped to bring the reports promptly and at frequent intervals in emergencies.

In this modern drama of fear and violence, the hurricane warning has become the signal that may cause desperate actions by hundreds of thousands of people. Colossal costs are entailed in the movement of populations in exposed places and in the protection of property and interruption of business. Now, in this emergency, a civil service employee not used to making decisions involving large sums of money finds himself in a position from which he has no escape. He has to make up his mind—to issue the warning or not to issue it. If he fails to get it out in time, there will be much loss of life and property that might have been avoided. If he issues the warning and the hurricane turns away from the coast or loses force, very large costs will have been entailed without apparent justification. In either case, he will be subjected to a lot of criticism.

The hurricane hunter and forecaster who stepped into this responsible position at a critical time was Grady Norton. Born in Alabama, in 1895, Grady joined the Weather Bureau shortly before World War I, then became a meteorologist in the Army, after taking training at A. & M. College of Texas, where a weather school was established early in 1918. But he had no wish to be a forecaster or to send out warnings of hurricanes.

Nevertheless, the people in Washington were unable to get out of their minds the fact that whenever Norton made forecasts for practice, his rating was very high, especially for the southeastern part of the country. The Bureau encouraged him at every opportunity because he was one of those who are born with the knack of making good weather predictions—which is an art rather than a science, even in its present stage of development.

Then in 1928, Grady went on a motor trip and arrived in southern Florida just after the Palm Beach hurricane had struck Lake Okeechobee, killing more than two thousand people. He saw the devastation, the mass burials, the suffering, and determined to do something about it. By 1930 he was at New Orleans, getting experience in forecasting Gulf hurricanes. After five years, the hurricane teletype and the centers at Jacksonville and New Orleans were established and Grady was put in charge of hurricane forecasting at Jacksonville. There, and later at Miami, his name, Grady Norton, coming over the radio, became familiar and reassuring to almost every householder in the region. For twenty hurricane seasons he took the brunt of it in almost countless emergencies. In some instances, he made broadcasts steadily and continuously every two hours, or oftener, for two days or more without rest, his microphone having direct connections to more than twenty Florida radio stations, and by powerful short-wave hook-ups to small towns all over the state. As the hurricane threatened areas beyond Florida, he continued the issue of bulletins, warnings, and advices. In the last ten years of this service, he was warned by his physicians to turn a good deal of the responsibility over to his assistants, but the public wanted to know his personal decisions.

In 1954, after Hurricanes Carol and Edna had devastated sections of the northeast with resultant serious criticism of the Bureau in regard to the former, a fast-moving blow that allowed very limited time for precautions, Norton died on the job while tracking Hurricane Hazel through the Caribbean. A tall, thin, sandy-haired Southerner, Norton had a slow, calm way of talking that put him, in the public mind, at the top of the list of hurricane hunters of his generation. And it was generally conceded that to his efforts were to be credited in a large degree the advances in hurricane forecasting in the years after 1935. But the outstanding progress was gained from the use of aircraft to reconnoiter hurricanes, in which Norton played a very important part.

In Grady Norton's place, the Bureau put Gordon Dunn, who was an associate of Norton's at Jacksonville when the service began and who had more recently been in charge of the forecast center at Chicago.

By the end of 1942 it was plain that the weather offices of the Army and Navy would have to join with the Weather Bureau in hunting and predicting hurricanes. It was agreed that the combined office would work best at Miami. For the 1943 storm season, the Weather Bureau moved its forecast office from Jacksonville to Miami, with Norton in charge, and the military agencies assigned liaison officers there for the purpose of coordinating the weather reports received and the warnings issued. All the experts felt that military aircraft would have to be used to get the reports needed. In August, 1943, the news of Colonel Duckworth's successful flight into the center of the Texas hurricane was the decisive factor. Reconnaissance began in 1944.

9. WINGS AGAINST THE WHIRLING BLASTS

>>

Said the black-browed hurricane
Brooding down the Spanish main
"Shall I see my forces, zounds!
Measured in square inches, pounds?
With detectives at my back
When I double on my track!
All my secret paths made clear!
Published to a hemisphere!
Shall I? Blow me, if I do!"

—Bret Harte

After Joe Duckworth flew into the center of the hurricane near Galveston on July 27, 1943, there was much excitement about the remarkable fact that he had experienced no very dangerous weather or damage to his plane on the trip. But the experts realized that hunting hurricanes as a regular business would be different. Men who had flown the weather in the Caribbean and elsewhere in the tropics and subtropics, and those who had just thought about it, had visions of undulating seas stirred by soft tropical breezes, white clouds piled in neat balls on the horizon, blue water, blue sky, and lush palm-covered coasts and islands. And yet they knew that nowhere is the sly trickery of wind and storm more dangerous. Suddenly and with no apparent reason, the soft breezes turn into quick little gusts and wrap themselves around a center, with gray clouds spreading and rain coming in brief squalls. The whirl spreads, gathering other winds into its orbit, and hard rain begins. Soon there are violent gales and the power of the storm is apparent in the roaring of the wind and sea.

And so it is easy to think of a plane in a hurricane as being like an oak leaf in a thunderstorm, except that the leaf is bigger in proportion but lacks the skillful handling of a youthful crew, alert, fearful and resourceful, straining desperately to keep it from rocketing steeply into the wind-torn sea below. For these reasons, the men who ventured in 1943 to probe tropical storms by air were exceedingly cautious about it. They went into it at a high level—usually as far up as the plane would go—and came down by easy stages, in the calm center, if possible, ready to turn around and dash for land the moment anything went wrong.

The next after Duckworth and his associates to look into a hurricane was Captain G. H. MacDougall of the Army Air Forces. The second fully-developed storm of 1943 came from far out in the open Atlantic and passed east of the Windward Islands on a north-northwest course toward Bermuda. MacDougall wanted to have a first-hand view of its insides. Ships in the Atlantic were reporting extremely high winds and waves fifty to sixty feet high and five hundred to six hundred feet in length. MacDougall went to see Colonel Alan, who said he was ready to pilot the plane. So the two took off from Antigua on August 20.

According to the report by MacDougall, they came in at a very high level and began to explore the outer circulation of the storm. He said: "We ran into rain falling from overcast.

There were broken cumulus and stratus clouds below us. As the sun became more and more blotted out, we seemed to be heading into a bluish twilight. In spite of the low visibility due both to rain and moderate haze, it was impossible to make out the ocean through the wind-torn stratus below, and while we were yet to see the teeth of the storm, the snarl was already too evident. A surface wind of forty to fifty miles per hour from the southwest was probably a good estimate in this part of the storm. Colonel Alan now began to let the plane down and we stopped taking oxygen. At the same time, the wheels were let down to minimize the turbulence, and the plane leveled off at an elevation of one thousand feet which was below the stratus.

"For those of us who had spent enough time in the Caribbean to be familiar with the magnitude of the waves usually encountered, it was hard to believe what we saw below. The seas were tremendous and the crests were being blown off in long swirls by a wind that must easily have exceeded seventy miles per hour. The long parallel streaks of foam streaming from one wave to another made it evident from which direction the wind was blowing."

About a month later, a tropical storm formed in the western Gulf of Mexico, not far from Vera Cruz. Shortly afterward, it moved toward the Texas coast, increasing rapidly in force, and there was general alarm. People began to abandon the beaches and protect their property in the coastal towns. At this time there was a young officer, Lieutenant Paul Ekern, at Tinker Air Field near Oklahoma City, who was anxious to see the inside of one of these big storms. This one looked like his last chance for 1943 and he began talking it up. He found Sergeant Jack Huennekens who was ready to go and they looked for a plane and pilot. Time went by, but the hurricane center slowed down to a crawl and described a loop off the coast, taking three days to turn around. Excited conversations about the storm created interest, and about the time that Ekern and Huennekens found an Air Force pilot, Captain Griffin, anxious to go, a Navy man came over from Norman, Oklahoma, and said he had some instruments he would like to carry into the hurricane and get records of conditions encountered. He was told that anybody crazy enough to go was welcome. He introduced himself as a Navy Aerologist, Gerald Finger, and they all shook hands and got their things ready.

On the afternoon of the eighteenth, with the hurricane still hanging ominously off the coast but with some loss of violence, the crew took off for south Texas, carrying the Navy man and his instruments. They came into the storm area at about thirty thousand feet and proceeded cautiously toward the center. At this level there was very little turbulence, but the view was magnificent. There were mountainous thunderclouds, some extending fifteen thousand feet above the plane. Carefully they explored the region and finally came into a place where they could see the surface of the Gulf white with foam and piled-up clouds ringing a space where the sky was partly clear. This, they decided, must be the center.

Cautiously they went down to twelve thousand feet, circling around as they descended, and keeping records of temperature, humidity, and pressure. At times they flew through clouds on instruments in the rain, and now and then there was light icing. After about three hours, they began to run low on gas, so they flew through the western part of the storm and back to Oklahoma.

At the end of the hurricane season, these flights were reported to the Weather Bureau and recommendations were forwarded to the Joint Chiefs of Staff that military aircraft be used routinely to explore hurricanes and improve the accuracy of the warnings. The Joint Chiefs referred this to their meteorological committee, with representatives of the Army Air Corps, the Navy and the Weather Bureau, and on February 15, 1944, a plan was approved for the coming season. As far as possible, crews with experience in flying the weather were selected. Some of these had been on daily missions on the Atlantic, for the protection of convoys. By the beginning of the 1944 season, planes and men were at their posts in Florida, ready to go on instructions from the joint hurricane center in Miami.

Probing of hurricanes by air came to a sharp focus in September, 1944. On the eighth, signs of a disturbance were picked up in the Atlantic, northeast of Puerto Rico. As it approached the northern Bahamas, its central pressure was extremely low, below 27.00 inches—estimated at 26.85—and it covered an enormous area with winds of terrific force. From here its center crossed the extreme eastern tip of North Carolina, sideswiped the New Jersey coast, doing vast damage, and then hit Long Island and New England with tremendous fury. On account of the war, ships at sea were not reporting the weather and the hurricane hunters had a real job on their hands.

On the morning of the tenth, Forecaster Norton at the Miami Weather Bureau studied the weather map, grumbled about the lack of observations from the West Indies, and decided to ask for a plane to go out and report the weather north of Puerto Rico. He had little to go on, but he thought it was a very bad storm. On the afternoon of the ninth, the Air Corps had sent a plane out from Antigua. They had reported winds of eighty miles, very rough seas, and center about 250 miles northeast of San Juan. Very little information had come from the area since that time, except the regular weather messages from San Juan. After trying to get the Navy office on the telephone half a dozen times, Norton gave up. Every time he started to dial, the phone rang and he answered it, making an effort to hang up quickly and get a call in before it rang again. But many people had learned about the storm and were anxious for more information, hence the phone was constantly busy.

"I thought this was an unlisted phone," he complained to the map crew. "It is," replied an assistant. "We gave the number only to the radio, press, and a few others, to make sure we could get a call out when we had to, but these restricted phone numbers leak out. We'll have to change the number again."

Norton squeezed between the map man and the wall and sat down at the teletypewriter in the corner after the operator had stepped out into the hall. The office was crowded and when one man wanted to leave his place, nearly everybody else had to stand up to make room. Norton rang a bell, rattled the teletypewriter, and finally got Commander Loveland on the line down at the Navy office.

This was an exclusive line—Weather Bureau to Navy—and Norton pecked out a message. "Looks like a bad hurricane out there. It's maybe three days from Florida if it comes here, but it probably won't. Looks like it would go up toward the Carolinas. We can't be sure. Maybe we should have a recco this morning. What do you think?"

"Think we can get one up there from Puerto Rico this morning," came the message from Loveland. "I'll see what I can do. Did you check with the Army?"

"Yes, the Major talked to Colonel Ellsworth and he says they expect to get a plane out there from Borinquen this afternoon. Also, I asked for clearance on a public message yesterday and got an OK last night."

At that time, because of the war, public releases about storms along the coast were still restricted and had to be cleared with Naval Operations in Washington. If enemy submarines learned that planes were being evacuated from airports on the seaboard, they were emboldened to come out in the open and attack shipping along the coast. Oil tankers and other ships would have a bad enough time in the storm without running into submarines openly on the prowl. But the Chiefs of Staff had to balance this against the possible loss of life and property in coastal communities.

On their mission to explore the storm, the Navy crew from Puerto Rico ran into heavy rain and turbulence. Visibility was nil as they approached the center. They stayed down low to keep a view of the ocean but found the altimeter badly in error. As soon as they broke out of the clouds, they found the sea was much closer than they had figured. The plane was almost completely out of control several times. They changed course, got out of the storm, sent a message to Miami, and returned to Ramey Field in Puerto Rico.

Steadily the hurricane kept on a west-northwest course, increasing in size and violence. As it went along, the aircraft of the Navy and Air Forces were on its heels and driving toward the center, like gnats around an angry bull. It was headed for the Carolinas; everybody was agreed on that now. Ships were in trouble, running to get from between the hurricane and the coast as the winds closed in, and anxious people waited for the next report.

At that time, a hurricane was thought to have four stages of existence. First was the formation stage, often with circulatory winds and rain developing in a pressure wave coming westward over the Atlantic or Caribbean. Second, it quickly concentrated into a small but very violent whirl and, over a relatively small area, had the most violent winds of its existence. In this stage it might not have been more than one hundred miles in diameter. Third, it became a mature storm, spreading out, and although its winds did not become any more violent, they spread over a much larger area, maybe as much as three hundred miles, or more, in diameter. Fourth was the stage of decay, when it began to lose its almost circular shape and the winds began to diminish. Now it went off to the northward and became an extra-tropical storm or struck inland in the south and died with torrential rains and squally winds.

This hurricane seemed to be an exception. As it spread out to cover a bigger area, its winds seemed to develop greater fury. A Navy plane went in as it approached the Carolinas and found extreme turbulence, winds estimated at 140 miles an hour, torrential rain that penetrated the airplane, and no visibility through the splatter and smear on the windows. And when the stalwart crew came down below the clouds, the sea was a welter of foam, with gusts wiping the tops off waves that reached up to tremendous heights.

While no planes were lost in probing this terrible storm, a destroyer, a mine sweeper, two Coast Guard cutters, and a light vessel were sunk. An Army plane estimated the winds at 140 miles an hour. The weather officer, Lieutenant Victor Klobucher, said that it was the worst storm that had been probed by the hurricane hunters up to that time. The turbulence was so bad that, with both the pilot and co-pilot straining every muscle, the plane could not be kept under control, and several times they thought it would be torn apart or crash into the sea. On returning to the base, the fliers found that 150 rivets had been sheared off one wing alone.

On the morning of the fourteenth of September, the terrible tempest was close to the eastern tip of North Carolina, apparently destined to sideswipe the coast from there northward with devastating force. There was some alarm in Washington. It might possibly turn more to the northward and its center might come up Chesapeake Bay or up the Potomac River. A violent storm in Washington at that time would have been detrimental to the prosecution of war plans. In 1933, a smaller hurricane had taken this course and its destructive visit to the Bay region and the Capital had not been forgotten. Also, in the minds of the military was the opportunity offered that day to explore a big hurricane and find out more concerning its inner workings.

On that critical morning, Colonel F. B. Wood, a veteran flyer in the Air Corps, came down to Bolling Field outside Washington with hurricane-probing on his mind. After talking about it to the men around the field, he decided to try a flight at least into the outer edges of the storm as it passed to the eastward during the day. He thought about the junior officers and men being sent into these furious winds and he felt it was a good idea for one of the head men to go out and see what it was like.

Wood talked to Lieutenant Frank Record and found he was anxious to go. He grabbed the telephone and got Major Harry Wexler on the line. Harry was a Weather Bureau research official who was in the Army for the duration.

"Harry, how about taking you out in the hurricane today?" Wood asked. "I'll pilot the plane. Frank is going along."

"Sure you can take me out, but you've got to bring me back," Harry answered. "This is a round trip, Floyd, I hope." Wood agreed to do his best to make a round trip out of it.

At two o'clock that afternoon, the trio took off and headed east with some misgivings. They knew that this was one of the worst tropical storms that had been charted up to that time. The hurricane was then centered near Cape Henry, Virginia. The wind at Norfolk had been up to ninety miles an hour. Colonel Wood described it as follows:

"Immediately after take-off, we penetrated a thin overcast, the top of which was about fifteen hundred feet, and then proceeded to a point approximately twenty miles northeast of Langley Field. The boundary of the hurricane, as seen from the latter location, was a dense black wall running along the western edge of the Chesapeake Bay. The airplane was turned on a heading so as to fly a track that would lead straight toward the estimated position of the center of the hurricane. Altitude was three thousand feet. A drift correction of 30° was allowed to account for the estimated one hundred miles per hour cross wind encountered at the outer edge of the storm. Immediately on entering the outer edge, the atmosphere turned very dark and a blanket of heavy rainfall was encountered."

Very surprisingly, the flyers reported that in this area a strong but steady down-current was also encountered. The latter was contrary to the accepted idea that all of the area encompassed by the steep pressure fall in a hurricane contains ascending rather than descending air up to great heights. Although visibility was very low, due to the heavy rainfall, there were very few clouds below the altitude of the airplane (three thousand feet), except for some scud over Cape Henry.

The waves in Chesapeake Bay were enormous. A freighter plowing through the Bay was being swept from bow to stern by huge waves which at times appeared to engulf the whole vessel at once. Spray was being thrown into the air at heights which appeared to reach two hundred feet above the surface of the Bay. From the appearance of the water, both within Chesapeake Bay and east of Cape Henry, it is not surprising that a Navy destroyer of the 1850-ton class was sunk there. One of the foremost thoughts in the men's minds at the time was that should the aircraft be forced down in the hurricane, neither life rafts, "Mae Wests," or any other livesaving device would have saved them from drowning!

The flight was continued on toward the assumed position of the center of the hurricane. Although the downdraft continued strong, very little turbulence was encountered. The airplane lost a speed of about seventy miles per hour in the necessary climb required to make up for the downward motion of the air. The heavy rain continued. At a point approximately fifty to sixty miles inward from the outer edge of the hurricane, they suddenly entered an area of rising air. This area also contained fairly dense clouds below, but very thin clouds above. The sun was visible through the thin clouds overhead. They seemed to be on the edge of the center. The vertical air movement was of such magnitude that the airplane was lifted from the three thousand foot level to five thousand feet before power could be reduced and the airplane nosed downward. Turbulence in this area was also considerably more severe than in the zone of descending air just passed through, but was not of such severity as to endanger the flight.

Although the flight was continued for a few minutes on toward the point where the center of the hurricane was thought to be, the conditions of flight remained constant; that is, moderate turbulence, rising air, and the sun faintly visible through the thin clouds overhead. The men thought they were off to one side or other of the center, but not finding it, and not knowing the direction in which to fly to locate it exactly, the airplane was turned around and flown on a track which was estimated would lead toward Norfolk. An altitude of five thousand feet was maintained on the way out. The dark band of descending air and heavy rainfall was traversed in the reverse order as during the incoming flight. They emerged from the hurricane at a point approximately thirty miles east-northeast of Norfolk.

Afterward, Colonel Wood felt more confident about junior officers flying into hurricanes, but there were many questions yet to be answered. Incidentally, the three men in this plane and the members of the squadron who flew into the same hurricane from Miami were awarded the Air Medal in February, 1945, for their bravery in these flights.

Colonel Wood drew the following conclusions:

"Although one of the more important points indicated by our experience during the aforementioned flight is that hurricanes can very probably be successfully flown through after they have reached temperate latitudes, it should not be accepted as conclusive proof that all hurricanes may be flown through. Although there have been several instances of flights into hurricanes before they migrated out of the tropical regions, it is not known whether, at the times the flights were made, any of these storms were of an intensity that even approached the maximum possible. Further, it is not known for certain whether the hurricane that passed along the Virginia coast on the fourteenth of September is typical of all hurricanes once they reach temperate latitudes. Indications are that this hurricane was about as severe as they ever get to be at these latitudes, but insufficient flying experience in hurricanes has been obtained to determine conclusively that all hurricanes in temperate latitudes are safe to fly through. Any pilot who in the future might desire to repeat the experience referred to in this statement is advised that any hurricane should be approached gingerly and with a view toward making an immediate 180° change in his track, should severe turbulence, hail, or severe thunderstorm activity be encountered.

"It is believed that the method of examining a hurricane by flight reconnaissance that would produce the most revealing results is to attempt an approach to it from the stratosphere. It is thought, further, that such a flight could be made over the outer rim of the hurricane and a let-down into the center or hollow eye of the storm be made with complete safety. A record of the temperature at various flight levels while descending through the central (hollow) portion of the storm, together with photographs of the cloud structure, would be of tremendous value."

In October there was another hurricane in Florida. It began in the western Caribbean on the thirteenth and crossed western Cuba on the seventeenth. On the south coast the hurricane winds created an enormous tide. More than three hundred people were killed, and a Standard Oil Company barge was carried ten miles inland. When the big winds roared across Florida on the eighteenth and nineteenth, it was a severe storm with a calm center that was at one time about seventy miles long.

As it drove violent winds and seas toward Florida, an airline company, Transcontinental & Western Air, decided to investigate and sent an experienced pilot, Captain Robert Buck, in a B-17, to fly through and observe the weather and electrical phenomena in the storm. Of course, he considered the flight hazardous but he was willing. Any person who had experienced the violent winds of these storms or read about their destructive effects was likely to assume that a plane at low levels in the middle part of the storm might have its wings torn off.

Buck started to climb into the edge of the storm at Alma, Georgia, going in warily at four thousand feet and finding only light to moderate turbulence from there to nine thousand feet, after which it became smoother. This was in accordance with the reports of other fliers who had ventured in at high levels, and he was reassured.

At eleven thousand feet the rain changed to sleet. This was not unexpected. Ordinarily it is much colder at such a height than at the ground. The temperature drops about one degree for each rise of three hundred feet. Although the plane was flying in instrument conditions and "blind," there were no ordinary water-cloud particles, but simply haze and sleet.

At 12,700 feet, with the temperature at the freezing point, the plane flew through moderate to heavy snow with very large flakes. The climb was continued and the snow remained moderate, but as the altitude increased, the size of the snowflakes decreased. The air was perfectly smooth, with the exception of about one minute of light turbulence at 16,000 feet. During the entire climb no ice was encountered, but there were a few patches of snow sticking on the airplane. This was definitely not ice. Due to loss of radio reception on all receivers, including the loop, it was difficult to obtain the wind accurately. It was estimated to be easterly at approximately eighty-five miles an hour, to about 16,000 feet, where it changed to westerly with about the same velocity.

At 19,400 feet, the temperature had dropped to 27°. At 22,800 feet, the snow was light and fine and the temperature was 18°. The temperature had dropped to 14° at 24,600 feet.

At 25,000 feet, the plane broke out of the side of the storm near the top. At 25,800 feet, the plane was flying in the clear where the temperature was 18°. During the entire climb from 9,500 feet to 25,000 feet, no fog was encountered, only particles of snow.

Near Jacksonville, Florida, the tops of the clouds dropped sharply to 8,000 feet. The plane flew east out to sea to check the eastern side of the storm and, satisfied that Jacksonville was close to the storm's center, proceeded to the coast again and to Daytona Beach, where the craft landed.

Pilot Buck concluded that the paramount danger lies in an aircraft becoming lost, due to the failure of radio navigation caused by static, coupled with the high winds. He said that a tropical storm of the type flown is not hazardous to aircraft in respect to structural failure and loss of control, if an altitude of over approximately 8,000 feet is held.

In December, all the men connected with the hurricane warning service in the Army, Navy, Weather Bureau and other agencies—including the top officials, the forecasters, the men who directed the flights, the pilots, weather officers, and others who made up the crews, the radio men on shore, and the Coast Guard people—were fully represented in a conference in Washington. Here they all went over their experiences and offered every possible suggestion for improving the service. Many things were needed, but two tough problems worried everybody.

One was how the crew could find out where they were in latitude and longitude or in distance and direction from some point on an island or on the coast after they found the center of the storm. After all, the weather observer, navigator, and the radio man might figure out how to get in the eye, and the plane might get into it, but if they failed to get their position accurately, the information was of doubtful value. This nearly always depended on radio signals from distant shore stations, for it was seldom that they could get a celestial fix as a mate does on a ship at sea. The second problem was communications—how to get the weather message off and be sure it had been received at a shore radio station, and see also that it reached the forecast offices promptly. All of this had many sources of delay. In a hurricane, the atmospherics were often excessive. At times the radio man on the plane could hear nothing but loud static in his ear phones. He was powerless to do anything except to send "blind" and hope somebody would receive it and understand what it was. Slowly these problems were solved in part as time went on.

10. KAPPLER'S HURRICANE

>>>

Black it stood as night,
Fierce as ten furies, terrible as hell.
—Milton

Kappler's Hurricane was one of the most violent of history. It got its name from a weather officer, a second lieutenant in the Army Air Corps named Bernard J. Kappler. The story includes the vivid personal reactions of a number of men who explored this tremendous storm as it built up its energy while crossing fifteen hundred miles of tropical and subtropical sea surface and finally ravaged parts of Southern Florida, including the outright destruction of the big Richmond Naval Air Base.

The fact is that this storm seems to have had its birth over western Africa. There were signs of it there and near the Cape Verde Islands on the first two days of September. Later there were some indications of its winds and low pressure in radio reports from ships but eventually it was lost for the time being, far out in the Atlantic.

Kappler discovered it on September 12, 1945. He was on a regular weather-reporting mission to the Windward

Islands. Every day one or more B-25's took off from Morrison Field at West Palm Beach and explored the atmosphere on flights to Antigua, British West Indies, returning via the open Atlantic to Florida. On that day there was nothing unusual until the plane in which Kappler was flying was about two hours from Antigua. Here, he noted a black wall of clouds to the east and at his suggestion the pilot, First Lieutenant D. A. Cassidy, took the plane down to fifteen hundred feet and they looked around.

Without any doubt, a tropical storm was in the making. Its winds already were blowing around a center with gusts at about seventy miles an hour. There was moderate turbulence, with stretches of rain, but they had no particular difficulty in flying through it. They reported it to headquarters and were told to land at Coolidge Field in Antigua and be prepared to take another look and report in the morning.

This operation was known as "Duck Fight," consisting of five B-25 aircraft and five crews made up of twenty officers and fifteen men. This particular group had been at British Guiana but had moved up to Florida in May for the new hurricane duty. It was their job to explore this region twice daily, looking for weather trouble when no storm was known to be in progress. If a suspicious area was found, they were deployed and used in accordance with directives from the hurricane center at Miami. The Navy also had planes assigned to similar missions.

After breakfast on the thirteenth Kappler's crew took off again. About two hours out of Antigua, they encountered winds up to about eighty knots (a little above ninety miles an hour) but flying was smooth. The crew made a few jokes on the general subject of how easy it was to fly through hurricanes. The co-pilot, Lieutenant Hugh Crowe, had the controls. He turned toward the center and the wind picked up to 120 knots. Soon they were in trouble, with severe turbulence and heavy rain. The air speed fluctuated between 160 and 240 miles an hour and cylinder temperatures began to fall rapidly. Crowe fed power to the engines, but the plane began getting out of control. Cassidy had to help him keep the ship level. Kappler shouted that the pressure was dropping rapidly—the pressure altitude was seventeen hundred feet but their actual height was only nine hundred. Crowe said the turbulence was the most severe he had ever experienced. The plane yawed fifteen degrees on either side of the heading. The navigator, Lieutenant Redding W. Bunting, said dryly, "In my opinion a hurricane is not the place in which to fly an airplane."

By the fourteenth, it was obvious to all concerned that they had a really big storm on their hands. Its center had been north of Puerto Rico on the thirteenth, and on the fourteenth, moving rather rapidly, it was passing north of Haiti. The first plane took off from Borinquen Field, Puerto Rico, in the morning, Cassidy at the controls, and within an hour the crew were getting into it. At the end of this flight, Co-pilot Crowe said, "My respect for hurricanes has increased tremendously!"

First, the right engine was not running smoothly and after a little it almost stopped. Cassidy asked Bunting where the nearest land was and when he said Cuba, they turned 90° and made for it. After twenty minutes the engine was doing better, so they had a brief conference and decided to try for the hurricane center. Turning back, they saw gigantic sea swells and a white boiling ocean ahead. Soon they hit the worst turbulence Cassidy had ever seen, and with it there were intervals of torrential rain. It was terrific. The cockpit was leaking like a sieve. Most of the time it took full rudder and aileron to lift a wing. The plane got into attitudes they had never dreamed of. It was impossible to hold a heading, for the ship was yawing more than 30° and taking a terrible side buffeting. Maybe this lasted three to five minutes but it seemed like hours. Suddenly they passed through the edge of the center, it was smooth for about a minute, and then they were in the worst part again. Bunting noted a piece of advice, "When you are near the center, about all you can do is brace yourself and hold on to something that won't pull loose."

Bunting reported afterward that it took both pilot and co-pilot to control the ship and at times the RPM set at 2,100 would drop to 1,900 and then rise to 2,200, due to the terrific force of the wind. Kappler kept phoning the correct altitude to the pilot at short intervals because of the enormous changes in pressure. It was impossible to write in the log book so he scribbled as best he could on a piece of paper and copied it afterward. He noted that before entering the eye it was very dark. Inside it was cloudy but the light was better, indicating that the upper clouds were missing. When the flight was finished the crew was glad to be back at Morrison Field—to put it mildly!

Another plane at Morrison Field had been out the day before and soon was taking off again, at 2:00 P.M. The pilot was Lieutenant A. D. Gunn. He flew a direct course to the center of the storm—he hadn't realized the day before that he was elected to go through it again today, so he wanted to get it over with as soon as possible. These two days had provided his first such experience. One cylinder head slid to a very low temperature in the heavy rain and Gunn dropped the landing gear and tried to keep it up to 100°, but one engine died. The turbulence was so bad that neither he nor the co-pilot could tell which engine was out. The severe turbulence lasted for a full thirty minutes, about ten minutes of this being flown on one engine, with the crew desperately working on the other while they bounced around. The flight engineer, Sergeant Harry Kiefaber, had to leave his seat because of water pouring down his back and the tossing up and down, with his head repeatedly hitting the top of the plane. He tried to go back to join the navigator but the plane started to fall off to the right and he had visions of ditching in a mass of white foam. The pilot got it under control but it seemed that they were being tossed around like popcorn in a popper. Gradually the turbulence ceased, the other engine began running smoothly and they headed straight for Morrison.

But the conditions on the fourteenth were just an introduction to what happened on the fifteenth. The first crew took off at 7 A.M., with the edge of the hurricane causing rough weather at the field. Here is the story told by the navigator, Lieutenant James P. Dalton:

"Frankly speaking, throughout my entire life I have been frightened, really frightened, only three times. All of this was connected intimately with weather reconnaissance. I think I can truthfully and without exaggeration say that absolutely the worst time was while I was flying through Kappler's Hurricane on September 15, 1945. We were stationed at Morrison Field, West Palm Beach, Florida, at the time. Everyone except the Duck Flight Recco Squadron had evacuated the field for safer areas the day before.

"Hurricane reconnaissance being our business, we of course stayed on, in order to operate as closely as possible to the storm. We were to take off at 7:00 A.M. local time and by then several thunderstorms had already appeared, thoroughly drenching us before we could climb into our plane. But each crew member was keenly alert, for he knew what to expect. I've flown approximately fifteen hundred normal weather reconnaissance hours; that is, if you can call going out and looking for trouble 'normal flying.' I have covered the Atlantic completely north of the equator to the

Arctic Circle, flying in all kinds of weather and during all seasons but never has anything like this happened to me before.

"One minute this plane, seemingly under control, would suddenly wrench itself free, throw itself into a vertical bank and head straight for the steaming white sea below. An instant later it was on the other wing, this time climbing with its nose down at an ungodly speed. To ditch would be disastrous. I stood on my hands as much as I did on my feet. Rain was so heavy it was as if we were flying through the sea like a submarine. Navigation was practically impossible. For not a minute could we say we were moving in any single direction—at one time I recorded twenty-eight degrees drift, two minutes later it was from the opposite direction almost as strong. But then taking a drift reading during the worst of it was out of the question. I was able to record a wind of 125 miles an hour, and I still don't know how it was possible, the air was so terribly rough. At one time, though, our pressure altimeter was indicating twenty-six hundred feet due to the drop in pressure, when we actually were at seven hundred feet. At this time the bottom fell out. I don't know how close we came to the sea but it was far too close to suit my fancy. Right then and there I prayed. I vouched if I could come out alive I would never fly again.

"By the time we reached the center of the storm I was sick, real sick, and terribly frightened, but our job was only half over. We still had to fly from the center out, which proved to be as bad, if not worse, than going in.

"Mind you, for the first time, and after flying over fifteen hundred hours, I was airsick; and I wasn't alone. Our radio operator spilt his cookies just before we reached the center.

"After a total of five hours we landed at Eglin, the entire crew much happier to be safely back on the ground. At the time of our take-off we really didn't think it possible to fly safely through a hurricane. Personally I still don't. And I say again, I hope never to be as frightened as the time I flew through Kappler's Hurricane. It isn't safe."

Lieutenant Gunn, the pilot who had been in it the day before, was a man who took things calmly. He reported his experience:

"This morning the storm was only an hour and a half from the field. The usual line of squalls around the edge of the storm was hitting Morrison Field about every hour and a half. Of course this trip was to take us through the very center.

"We left Morrison at one thousand feet. The entire flight was turbulent and rainy. We circled the storm counterclockwise again and ran into the same turbulence and rain as before. This time the clouds must have been as low as five or six hundred feet, as even though we were only at one thousand feet, we could seldom get a glimpse of the ocean, which was churned up to such an extent that it seemed to be one big white cap. The altimeter was off one thousand feet at one point placing us at five hundred feet; then we could see the water. I believe even the fish drowned that day. As we entered the northeast quadrant, it got so rough that both pilot and co-pilot were flying the ship at the same time. The winds were so great at this point one could actually see the ship drifting over the sea. I think we had a drift correction of thirty-five to forty degrees at times.

"I don't think anyone will form a habit of this particular job. Prior to taking off I tried to take out hurricane insurance but it seems that they have no policies covering B-25 planes. Anyway, all the insurance salesmen had evacuated to some distant place like Long Beach, Calif."

Sergeant Robert Matzke, the radio operator, put it this way:

"September 15 was the day that I was picked on a crew

to fly the hurricane. Having been forewarned by several of the boys who had returned from the hurricane the day before, I set myself for something a little rougher than a weather mission with occasional turbulence. I figured that we had flown through what could well be considered rough weather while flying reconnaissance out of the Azores and maybe the boys were trying to throw a little scare into us as new men to the Morrison initiation.

"It seems that we had no sooner left the ground when we encountered rain and turbulence. This made me sort of leery of what was to come and I figured that if I were to send weather messages while in a hurricane, I'd have to send blind as the receivers were noisy already, and to hear and answer to a call would be almost an impossibility. As we proceeded toward the storm the rain became more intense and things were getting quite 'damp' in the ship. There was a leak right over my table and the steady downpour of water through this opening made it necessary for me to write with the log tablet braced against my knee to keep it from getting wet.

"The awful bouncing was getting my stomach and when we actually entered the hurricane it took all my strength to reach for the key to send a message. After a while I called to Lieutenant Schudel, our weather observer, and told him that I was sick and would have to rest my head on the table for a while. I had felt bad in a plane before but this was the first time that I was deathly sick. After a few minutes it was with all the strength that I could muster that I rolled my head to one side of the table and lost a few cookies.

"After I vomited a while I felt one hundred per cent better and I went to work pounding out the messages that had accumulated. It was impossible for me to hear any signals on the receivers due to atmospherics, so I sent blind, repeating myself over and over, in the hopes that someone would copy and relay to Miami for me. Our ships were vacating to Eglin Field that day and Sergeant Le Captain was standing watch on the frequency I was using. He came through with a receipt when I got to where I could hear in my receivers again.

"The flight that day was the roughest I have ever been on and a lot of my time was taken up just holding on for dear life and watching the B-4 bags bouncing up and down en masse like a big rubber ball. I was glad when the wheels hit the runway at Eglin Field and hungry, too, for my breakfast had stayed with me for a very short time. I imagine I looked rather beat up when I stepped from the plane but the ground felt so darn good under my feet and I didn't care who knew that I had been sicker than a dog."

Each member of the crew saw a little different part of the picture. Boys who flew these missions regularly became matter-of-fact in their reports and it was only when they were involved in a really big storm that they talked frankly about their feelings. Here is the story of the flight engineer, Sergeant Don Smith, in Kappler's Hurricane on September 15:

"The morning of the fifteenth loomed dark and formidable. This was our day to take a fling at the hurricane the other boys were telling us so much about. As a matter of fact it doesn't make you feel as though you were going on a Sunday School picnic. From the time we took off until we hit the storm we encountered turbulence and white caps were dashing around like mad but they were mild compared to what was coming.

"We circled the storm before heading for the center. We were hitting rain and moderate turbulence all this time. All at once we broke through the overcast and for a few seconds I wondered if it were letting up, but only for a second. One instant everything was peaceful and the next instant we were getting slapped around like a punching bag with Joe Louis on the prod. I looked at the bank and turn indicator and the rate of climb, and they both looked as if they were going all out to win a jitterbug contest. Now it was really raining. You've never seen it rain until you've been in a hurricane. I couldn't even see the engines from the cockpit window. I knew our right engine was the least bit rough before we started out and all I could think of was 'For gosh sakes, don't be cutting out now.' Before we were out of it, the engine sounded like a one-cylinder Harley motorcycle but really she never missed a beat. It was about this time that our cylinder head temperature dropped down to about 90° and the pilot dropped the wheels to bring it back up. And it was also about this time that we started for a milder climate.

"Don't ask me if I was scared or not. It would only be a fool or a liar who would say he wasn't worried. One thing about it is that you're so busy hanging on and trying to keep from getting thrown on your face that there isn't much time to think whether you're scared or not. It's really rough but there are no words to describe it. You'd have to go along to get the picture."

Lieutenant Kappler, for whom the hurricane was named, was due to go to Eglin Field with the crew that penetrated the hurricane on the fourteenth, but he wanted to stay over and see more of it. So they took him on, and although they already had a weather officer, Lieutenant Howard Schudel, Kappler was allowed to go as photographer. Schudel made the weather report from which the following is extracted:

"The rain was moderate at a distance from the center but already I was drenched because of a leaky nose in the ship. We flew almost completely around the center with nothing especially spectacular. At about twenty miles from the center we encountered severe turbulence which lasted only until the center was hit. During this time is when I found myself trying to code two weather messages at once and not doing a very good job on either. I actually was too busy to get very scared as to whether or not the plane would hold together. Between the severe turbulence and the water which by then had covered the entire desk, I could hardly read my own writing a half hour later when I was able to send the messages to the radio man. The turbulence near the center was of a nature I had never experienced previously. It was not a sharp jolt as experienced in a cumulus cloud but more of a rhythmic up and down motion. But on top of this there was a motion from side to side that made it especially rough.

"To me the most unwelcome sight of the whole trip was the swelling, churning sea. From nine hundred feet, which seemed to be our average altitude, the height of the spray above the ocean could not be determined. In places the surface was covered with sharp white streaks. If one thought for very long about what would happen to him if he were forced down upon this boiling ocean, he would be cured of hurricane flying for some time to come.

"The center was very welcome. The turbulence there was only light and the intense rain stopped completely. This gave me a momentary 'breather' so that I could swallow my stomach, assure myself that I was not sick, and code up a few back messages."

The morning crew went to Eglin Field and only one ship and crew was left at Morrison as the big storm closed in. The weather officer on this last flight was Lieutenant Edward Bourdet. He said:

"The weather during the entire morning at Morrison was bad. There were numerous thunderstorms with heavy rain showers that reduced visibility at times to less than one-quarter mile. Our flight took off at 10 A.M. We went just east of Miami where the wind was easterly at about fifty knots. We circled the storm center according to instructions and the wind went around from east to north and then through west to south. We experienced not only vertical currents but shearing horizontal currents. It is surprising that an airplane can hold together under such punishment. I found that there is no dry place in the nose of a B-25 in hurricane rain and I had to sit on the papers to keep them fairly dry, but I was also troubled in trying to keep myself from being battered against the side of the plane. We did not enter the eye of the storm but were in the northeast corner. The pilot later remarked, 'Our left wing tip may have been in the calm, but we sure as hell weren't.' It was here that I experienced the worst turbulence and the heaviest rain I have ever seen. The noise was terrific."

Lieutenant Bourdet added:

"The worst part of flying hurricanes is the fact that if there should be some trouble, structural or otherwise, that would force the plane down, the crew would not have a chance of getting out alive. The best part is the fact that you know that you are instrumental in providing adequate warning to all concerned and in saving lives and property."

During the time when these crews were flying into Kappler's Hurricane and sending reports to the Miami center, on September 15, the people of Florida were making last-minute preparations. Windows were boarded up, streams of refugees filled the highways, the radios were full of warnings, and the venturesome stood on the street corners as the gales began roaring in the wires and big waves came booming against the coast. Palm trees bent nearly double and debris began to fill the air. There was great damage at the Richmond Naval Air Base. Three big lighter-than-air hangars were destroyed. They collapsed in the wind at or near the peak of the hurricane and intense fires, fed by high octane gasoline, consumed the remains.

An investigating committee found that the winds must not have been less than 161 miles an hour to account for the bending of the large steel doors. Weather records recovered from the base indicated a two-minute wind of more than 170 miles an hour and as high as 198 miles an hour for a few seconds.

The center of the hurricane crossed the southern tip of Florida and moved up the west coast on the sixteenth as it turned north-northeastward, and then swept over Georgia and the Carolinas. Its center lay on the Georgia coast on the seventeenth. The boys who flew to Eglin Field had to take it again as its center came near and some of them flew into the hurricane after it passed Eglin. Among these was another weather officer, Lieutenant George Gray, who had seen this storm in several different places and now viewed it from the air as it whipped the Georgia coast. His report is worth reading:

"Riding through 'Kappler's Hurricane' was as rough a trip as I ever care to take. Admittedly, I know very little about flying from a pilot's point of view—how hard it is to keep a ship steady, the gyro, the cylinder head temperature, and all the rest that had the boys so worried. My criterion for roughness has always been how hard it is for me to hold on and how much the air speed fluctuates. We up front had to hold on with both hands when the going got bad. Some of the boys in back, we heard, with close to a thousand hours reconnaissance flying, actually got sick. The thing, though, that really frightened us was not the turbulence so much, because we had had to hold on with both hands before—it was the rain and the white sea below us.

"We saw the rain first from aloft. It looked absolutely black, as if a sudden darkness had set in in that part of the sky. The blackness seemed to hang straight down like a thick dark curtain from a solid altostratus deck at about fifteen thousand feet. How much further above this layer the build-up extended, I do not know. I kept thinking, 'We're not actually going into that.' We did though, and somehow with all the rush, we didn't have so much time to worry and become frightened as we expected. The rain was really terrific. It leaked in the nose and ran in a flood down the crawlway. The nose usually leaks and a soaking on a trip is not at all unusual, but this was different. I have never seen the water pour in and spurt so before. Where the plexiglass meets the floor section there was a regular fountain about six inches high that flooded the whole area. The noise was terrific. It pounded and crushed against the top and sides till we thought it would all collapse in upon us. I didn't notice any particular temperature change in the heavy rain though the pilots afterward all reported enormous cooling in the engines. Writing was almost impossible. The forms and charts on the table were like so much papier-mâché. There was no place that we could put them out of the water's way.

"We noticed the ocean particularly on the last day when the storm swept out to sea again off the Georgia coast. The day before on our way back to Morrison Field from Eglin where we rode out the blow, we flew low over the Everglades and saw roofless homes and millions of uprooted palmettos. The next day as we flew up the coast, we could see other remnants of the storm—huge pieces of timber, trees, roofs of outbuildings, and maybe even houses. The interphone was busy all the while as first one and then another of the crew saw something also afloat. As we got nearer the storm but still only in the scattered stratocumulus which is typical of almost any over-water flight, the rubbish seemed to disappear. Whether it was simply that the water itself was too rough for the timber to stand out or whether everything lay below the seething whiteness, I don't know. On our first trip into a tropical storm, the navigator kept repeating over the interphone, 'That water gives me the creeps.' It did. I kept thinking about ditching in it and floundering around in a 'Mae West'; I guess we all did. The waves were huge. Every now and then one would crest up and just as it was about to crash, the wind would grab hold of the foam and mist and crash it back into the sea. I took several pictures of the gradually heightening sea, though I doubt that its seething, alive look could be transposed to paper.

"We saw the storm hit the Carolina Cape. It was easy to see how trees in the Florida swamps without much root to grasp the earth were uprooted. Trees along the Carolina and Georgia coasts—big ones, taller than the houses in the vicinity—were bending before the blow the way wheat seems to ebb and flow in a summer's breeze. The seas were very high and in occasional breaks in the lower clouds we could catch glimpses of yellowish breakers and a littered beach. It looked as if the rain and thrashing surf had churned up the bottom, and mud had mixed with the foamy water. The shore was littered with debris, big trees, and blackened seaweed, mostly. As a sort of aside, on the matter of stirring up the bottom, we found several conch shells and bits of coral on the beach after the storm that are not considered native in these parts.

"Whether this next is typical of hurricanes or merely evidence that the storm had spent itself, I don't know, but I do think it worthy of mention. We noticed occasional breakups in the clouds—not large areas, just a few seconds when everything brightened and when the firm outlines of a large cumulus could be seen through thin low scud. This was not in the center but as much as forty miles away where the stuff should have been most solid and where the sea was near its roughest. I have seen the 'Eye' of a hurricane on land as a weather forecaster. At that time we noticed a real breakup with stars and moonlight visible. The wind and noise stopped for a while and we could see an occasional bulging cumulus through the night. Whether this phenomenon is due merely to less energy available over land than over water, I wouldn't even guess. In any event we noticed no such complete break in the eye at sea. In the center, so-called calm, though for my money it was mighty rough, about all that we noticed was that the pounding rain stopped for a minute or so. The clouds did not break clear through. There was a slight breakup to perhaps five thousand feet. There were bases of cumulus and several indefinite layers below this overcast though. The terrific bouncing around also stopped. We were out of the place in just a minute or two, so the eye couldn't have been much more than five miles in diameter. Some of the other ships circled in the center, saw a flock of birds milling around there, and noted violent up and down drafts near its edge. We were in and out of the thing so fast that, frankly, we hardly had time to notice anything. I think we could have fallen the seven hundred feet to the water without my knowing it, we were so busy with the camera, papers, and instruments.

"I might say a little more about the cloud formations we noticed since it was my job on this day to note them and take pictures of them while the other observer tried to compute pressure. Ahead of the storm here at Morrison Field on the morning of the sixteenth, we got a good picture of pre-hurricane thunderstorms. Squalls with forty-mile gusts swept across the runways. The rain came down in sheets so that we could watch it move toward us like a dark wall. Some of the boys out loading one of the ships for evacuation saw one of these terrific showers bearing down on them and they started to run for cover. The water was moving faster than they could run and before they'd moved fifty feet they were soaked to the skin. On the morning of the seventeenth, it lay just off the Georgia coast and had started to re-deepen. We flew up the eightieth meridian though it was hard to hold any steady course. As some of the navigators have probably mentioned, we could see our own drift. After we noted a good windshift into the east to assure us that we were in the northeast quadrant, we headed across current for the center and once there headed roughly for the great outside to the west. With such terrific drift, I don't see how anyone knew where he was going.

"Heading north: The usual over-water five-tenths stratocumulus bases at two thousand, tops at thirty-five hundred, gradually began to lower at about one hundred twenty-five miles from the center to roughly eight hundred feet, and a fairly solid lower layer of clouds. Flying above this layer at about forty-five hundred feet we could see tall bulging cumulus and thickening altostratus at about fifteen thousand ahead. There were other thin layers of stratocumulus and altostratus, but it wasn't until we got within fifty miles or so of the center and the rain really began to come down and the cumulus were as thick as trees in a forest that these intermediary layers began to thicken and thatch in between the tall cumulus the way they do in any well-developed storm system. By fifty miles out we were in solid cloud and heavy rain. Picture-taking became impossible except in the occasional breaks mentioned above. Even these breaks, if they should come out, would show little because continuous instrument weather, to me at least, looks pretty much the same whether it's part of a violent hurricane or smooth circulation stratus over a seaboard town. You can see the wing tips and not much more.

"If a general conclusion is necessary, mine would simply be that I'd just as soon not tempt fate in any more such storms."

Sometimes birds such as Lieutenant Gray describes are carried hundreds of miles before they escape from the hurricane. Species from Florida have been found as far north as New England.

11. TRICKS OF THE TRADE

A gallant barque with magic virtue graced,
Swift at our will with every wind to fly;
So that no changes of the shifting sky,
No stormy terrors of the watery waste,
Might bar our course,

—Dante

After two years of probing tropical storms by air, nearly everybody connected with the operation agreed that it was hazardous. But most of the men who were active in it had one main idea. As soon as the winds, rain, clouds, seas, and calm center of the average hurricane had been thoroughly mapped, a standard method should be devised for flying into the center and getting the vitally needed weather information en route with the least possible danger to the craft and crew. They thought of something like a football team, each man highly trained in a definite job, with faultless teamwork, and all members of the crew on the alert every moment.

Courses of instruction were organized. In all of them one fact became abundantly clear in the first two years. No two hurricanes are exactly alike. All of them are big compared with thunderstorms and tornadoes, but some are much larger than others. The recco crew may run into one in the uncertain stages of formation and at other times they may be nosing into an old storm with strange and unsymmetrical parts. Of certain elements they were reasonably sure—all these storms have clouds, rain, squalls, and central low pressure, with strong winds spiraling more or less regularly in a direction against the motions of the hands of a clock.

With these thoughts in mind, the instructors tried to devise methods that would prevent accidents. "What do you mean, accidents?" asked a junior weather officer at one of the conferences. "The whole thing is just one big accident, if you ask me. There's only one rule that's any good. Just be careful and don't fall in the ocean!" As a matter of fact, most of the rules had that one vital thought in mind, but there were different ways of doing it.

The Air Corps and Navy soon developed their own special methods. From the beginning the Navy preferred the low-level method; that is, they flew by the quickest route to the calm center of the storm, going in at a low level, generally at an elevation between three hundred and seven hundred feet. There are good reasons for this. Weather information—especially the facts they want about tropical storms—is vital to the safe operation of surface ships such as cruisers, destroyers and mine sweepers, and it is also used in the movement of aircraft from and to the decks of carriers. Task forces want to know about the speed and direction of winds at sea level, as well as the condition of the sea when storms are imminent.

It was the aim of the Navy to keep their weather reconnaissance aircraft below the level of clouds, where the aerologist could watch the surface of the sea as much of the time as is possible within the limits of reasonably safe operation.

When in a tropical storm, the aerologist guided the pilot around or into the center. Down near the water, say one hundred to three hundred feet altitude, turbulence is apt to be very bad, sometimes extremely violent. Above seven hundred feet, clouds are likely to interfere and this was extremely dangerous at that altitude in those early years because the altimeter which they used to indicate height of the aircraft by pressure of the atmosphere was sometimes badly in error in a tropical storm. If the pilot and the aerologist lost sight of the water's surface for a few minutes, they suddenly found the aircraft about to strike the precipitous waves of a storm-lashed sea.

Pressure of the atmosphere falls with increase of elevation, roughly one inch drop in pressure for each one thousand feet. If we put an ordinary barometer reading 29.90 inches in a plane on the ground and go up one thousand feet, it will read about 28.90 inches. The pressure altimeter is a special type of barometer that shows elevation instead of pressure. When the pressure is 29.90 inches and the altimeter is set at 0, we go up to where the pressure is 28.90 inches and it reads one thousand feet. But if the pressure over the region falls to 28.90 inches and the altimeter is not adjusted, it will read one thousand feet at the ground and be roughly one thousand feet in error when we go up to where the reading is 27.90 inches.

In ordinary weather, big changes in the barometer take place slowly and there usually is plenty of time for correction. In a flight into a hurricane, big changes take place rapidly. The change caused by the plane going up may be confused with the drop in pressure in the hurricane. If the plane is in the clouds when these changes take place, the pilot may have a frightening surprise on coming into the clear again. More recently, the hunters have been equipped with radar altimeters which give the absolute altitude for check. They send a radar pulse downward and it is bounced back from the sea surface to the instrument. The time it takes to go down and back depends on the height—the higher, the longer it takes—and the instrument is designed to give the indication very accurately in feet. Thus, the radar altimeter removed some of the dangers of low level flight.

So the Navy hunters moved in at low levels, preventing the "mush from becoming a splash" as they put it, and although their experienced pilots were marvelously efficient in flying on instruments in clouds or "on the gauges," they kept the white welter of the storm-lashed sea in view whenever possible. Of course, it is not possible to fly straight into a storm center. The big winds carry the plane with them and so the pilot might as well use the winds to good advantage—he will go with them to some extent, whether he likes it or not.

If we imagine ourselves in the center of the hurricane, facing forward along the line of motion of the storm itself—not the motion of the winds around the center—we know that the safest sector to fly in is behind us on our left, and the worst is in front of us on our right. At the left rear, there is likely to be better weather—less dense cloudiness and not so much rain. The winds are not so violent. So the Navy pilot flies with the wind. He goes in until he has winds of, say, sixty miles an hour. He puts the wind on the port quarter and this carries him gradually toward the center of the hurricane.

When he gets the wind speed to suit him, he brings the wind between the starboard quarter and dead astern and flies ahead to the point where he thinks he has the best place to go for the center. According to Commander N. Brango, one of the Navy's top specialists in hurricane navigation by air, "Choosing the proper run-in spot is tricky business, for it is the point at which the wind is the reciprocal of the storm's direction of motion. The pilot must watch for this point carefully, as he may pass it quickly; if he does there is imminent danger that the drift may carry the aircraft into the most severe quadrant of the hurricane." So the pilot goes into the center without wasting any time. Delay results in fatigue and it is important that the men be freshly alert. The pilot puts the wind broad on the port beam and he cannot possibly miss the eye. The next thing, the plane is in that amazing region where the sea boils, the breezes are light or missing altogether, the rain has ceased and the clouds are arranged in circular tiers, like giant spectators in a colossal football stadium.

This is a marvelous place. The crew is at ease. Coffee goes around. In the last few moments before coming into the eye, the craft leaks like a sieve. Everything is wet but the squirting from a hundred crevices in the plane ceases in the center and now it is possible to do some paper work. The aerologist is busy with the weather code and the radio man begins pounding out a message. They circle around. The pilot takes them up to maybe five thousand feet altitude and back down again, circling around.

And then the time comes to leave the center. The pilot calls a warning over the phone and there are two or three wisecracks. But this departure from the eye is dangerous. The plane begins to catch the shear of powerful winds around the center. Here a man can get thrown around violently and be seriously hurt, if he fails to get a good grip on something or neglects his safety belt.

Now the pilot sets the wind broad on the starboard beam and both he and the co-pilot hang onto the controls. This is rough going and there may be some surprises, but after a little they are out of the big wind circle and the navigator thinks the gales are down to something like fifty knots. The pilot sets course for the Navy airfield and the staccato notes of the radio continue to carry vital weather information to the forecasters. On this subject, Captain Robert Minter, an old hand, at one time in charge of aerology in the Office of Naval Operations, is full of enthusiasm. He guaranteed that the Navy could get a ship off the ground on a hurricane probe within an hour after the Weather Bureau forecaster asked for the information.

The Air Force has a different problem. Like the Navy, they are dedicated to the task of getting vital weather data for the forecasters, but their own problem is to evacuate military aircraft from threatened bases and get information needed for aeronautics. Also, they have the responsibility of giving weather forecasts and warnings to the Army. Until a few years after World War II, the Air Corps was a part of the Army, and when all three services were joined in the Department of Defense, the Air Force kept the weather job for both departments as a matter of economy and efficiency. Therefore, for this and other reasons, the Air Force follows a hurricane-probing plan which differs from the Navy's.

Flying generally at higher levels in tropical storms, the Air Force, as much as the Navy, puts a great deal of reliance on radar, which has become a marvelous aid in watching the weather. In the beginning—years ago—radar was not designed for weather purposes, however. During World War II, radar was used to spy on enemy ships and aircraft in fog or in darkness, to distances of 150 miles or more. The high-frequency rays sent out by the radar strike the object and are reflected back to the transmitter, where a sort of a silhouette appears on a scope. It may be black with white areas showing images of solid objects, such as planes and ships. In those days early in World War II, the weather was a nuisance to the radar people. It often seemed to interfere with the use of radar for military purposes, but the operators soon learned that the interference came from rain drops in local or general storms and that the rainy areas could be located and followed on the scope and, with the proper design, the apparatus could be used as a weather radar.

The first experiments with radar carried on board aircraft in organized tropical storm reconnaissance were made in 1945. Within three years, all the planes were carrying radar sets and had crew members whose sole business it was to watch the radar scope and tell the pilots and weather officers what kind of weather lay ahead.

Scarcely had these observations begun when the radar weather men discovered an amazing fact. On the radar, a tropical storm looks like an octopus with a doughnut for a body and arms that spiral around the body as if the creature had been caught in a whirlpool. These arms are bands of squally weather, oftentimes violent turmoil. Between the bands (or octopus arms) the wind is furious, of course, but there is less turbulence and cloudiness, and here the aircraft is in much less trouble than in the squall bands. The cause of these violent bands spiraling around the center has not been figured out yet for sure, but all tropical storms have them, and the hunters are beginning to understand them better.

The distance you can see from the radar station depends on how much weather there is. If there are large patches of dense rain, they may reflect all the rays back to the receiver and none may go through to show other rain areas farther away. Because of this, the radar shows the eye of the storm, but usually not the entire circle of clouds around a distant eye. Not enough radar energy is left to reflect from the opposite side of the eye. For this and other reasons it is necessary to have an experienced man to interpret the images on the radar scope.

From a radar in an airplane at high levels, these limitations are not so troublesome. Recently, too, the range of military radars has been increased. Whereas the radar formerly was very useful in getting a view of the eye from the aircraft, it did not give the eye's geographical position, which had to be determined by other means, except when the eye was close enough to be seen from the coast. With increased range, the aircraft can get between the hurricane center and the coast or an island, and both appear on opposite sides of the radarscope. In such cases, the distance and direction of the eye from a known point on a coast or island can be figured.

In the last two years, the Navy has used radar methods of this type extensively to obtain fixes of hurricane centers at night. In these instances, the crews fly at greater heights than in daylight and can get the eye and the coast on the scope at the same time. This gives a good estimate of center location to supplement the daylight penetrations without flying into the storm center in darkness. Actually, night flights directly into hurricane centers were not profitable, as non-radar observations of sea surface, clouds and winds were not possible in darkness.

It is apparent that a plane going into a storm at some upper level soon gets into the clouds and the sea surface is no longer visible. But the crew can depend on the radar to help find the center and they can go down in the eye of the storm and look around and, if necessary, the plane can descend in the outer parts of the storm and get estimates of the wind by a drift meter. For this latter procedure, the Air Forces at one time used what they called a "low-level boxing procedure." On this we can get the facts from the instructions issued by the head of the Air Weather Service, Brigadier General Thomas Moorman, Jr., a veteran of weather operations in World War II and in charge of weather reconnaissance in the Pacific, including the work done so effectively during the Korean War.

In 1953, Moorman directed that, in the interest of flying safety, there will be no low-level penetration of hurricanes. The Air Force pilots were asked to go into and out of the eye at the pressure level of seven hundred millibars which, under average conditions, is at about ten thousand feet altitude. Within 100 miles of a land mass, the flights in a hurricane would be at a minimum altitude of two thousand feet. To put it, in part, in the General's words, the hurricane mission would be conducted as follows:

For high-level penetration, the first priority would be given to obtaining an observed position of the storm center, either by a radar fix plus a navigation fix on the aircraft position, or a position found by penetrating the storm and obtaining a navigation fix in the eye. The storm would be approached on a track leading directly toward the center. If the storm center could not be reached at the seven hundred millibar level, the low-level boxing procedure could be followed, but if the radar set was not operating, no attempt would be made under these conditions to go into the eye.

For the low-level boxing procedure, the following instructions applied, quoting General Moorman in part:

"The storm area is approached on a track leading directly to the storm center and may be approached from any direction. As the winds increase in velocity, corrections will be made so that the wind is from the left and perpendicular to the track. The point at which the box is started is the midpoint of the base side of the rectangular pattern to be flown around the storm. When winds of sixty knots are encountered, the first leg will be started with a 90° turn to the right.

"The low-level box will be flown within the 45-60 knot wind area maintaining a true track for the first half of the leg, then a true heading for the succeeding legs. Surface winds should be 45° from the right when the left turn is made to the next leg. Double driftwinds should be obtained on each corner observation and each mid-point when practical. Reconnaissance of an area of a suspected hurricane will be flown with the same procedure.

"The weather observer will check the co-pilot's altimeter at frequent intervals to insure that it is reading the same as the radar altimeter.

"All flights will depart storm area prior to sunset, regardless of the degree of completion of the mission.

"Flight altitude while boxing the storm will be a minimum of five hundred feet absolute altitude, or at such higher altitude as will permit observations of the sea surface without hazard to safety. If contact flight cannot be maintained at five hundred feet, the legs will be flown a greater distance from the eye."

The "boxing procedure" was used a great deal by the Air Weather Service in the early years but by 1954 it had been eliminated. The seven-hundred-millibar method was revised, and as used in flights out of Bermuda in 1954 was described by Captain Ed Vrable, navigator, in part as follows: "(1) The aircraft flies down wind at right angles to the storm path to a point of lowest pressure, about twenty miles directly in front of the eye; (2) Flight is continued down wind for three minutes beyond the low point and then the heading of the aircraft is changed 135° to the left; (3) The aircraft continues on this course until the pressure begins to rise and then turns 90° to the left and into the center."

This new Air Force plan of flying into the hurricane at seven hundred millibars (ten thousand feet, roughly) is much like the Navy's low-level method, except that the Air Force crews enter down wind across the front of the storm, but this is nearly always an advantage for aircraft based at

Bermuda. From that island their most direct approach to an oncoming storm is into the front semicircle.

The Air Force has another aid in measuring weather in a storm. It is an instrument called a "dropsonde," a specially designed apparatus which works on the same principle as the older "radiosonde." A marvelously ingenious instrument, the radiosonde is a unit of very small weight containing miniature instruments for measuring pressure, temperature and humidity. It also has a metering device, a battery, and a small radio transmitter. The apparatus is carried aloft by a rubber balloon filled with helium. As the balloon rises, the radio transmitter sends signals for pressure, temperature and humidity at each level reached, and the signals are copied on a register at the ground weather station.

The dropsonde is a radiosonde that is thrown out of the aircraft flying at a high level, and allowed to descend by parachute, instead of being carried up by a balloon. There is a special listening post in the plane, where the data are recorded as the apparatus descends. The data are then put into the form of a message for transmission by the plane's radio operator to the forecasting base. This work with the dropsonde is usually done by the radar operator, in addition to his other duties.

Much of this fascinating work is done by the Air Weather Service of the Air Force on routine daily flights, whether or not there is a tropical storm to be studied. As an example, they have made daily flights from Alaska to the North Pole and back, to keep tabs on the strange weather up there. In this way, there—and in other parts of the world—they get weather daily from places on land and sea where there are no weather stations, no merchant ships to report, and no people to act as weather observers. These flights are named after some bird common to the region. The North Pole flight is called "Ptarmigan"; others are called "Vulture," "Gull," etc. Special flights into tropical storms in the Atlantic and Caribbean are called "Duck" missions.

Some of these improvements in the hurricane-hunting methods of the Air Weather Service were mentioned in a report by Robert Simpson, a Weather Bureau meteorologist, who flew with the Air Force into "Hurricane George" in 1947. This was a big storm which appeared first over the ocean to the eastward of the Lesser Antilles. The squadron assigned to the job had been moved to Kindley Field, at Bermuda. Simpson saw Lieutenant Colonel Robert David, who was in command, and arranged for the flight in one of the new planes piloted by an experienced officer, Lieutenant Mack Eastburn.

Hurricane George, so-called by the Air Force boys, although such names were not then official, moved slowly and menacingly across the Atlantic, north of Puerto Rico, and headed toward Florida. Simpson was in it several times with the Air Force. On the first flight, they were in an old B-29 which had too many hours on the engines and had been a bad actor on previous missions, but this time it behaved like a lady and they picked up a great deal of useful information. On the next trip they had a new plane. Here is a part of Simpson's story:

"Success is a marvelous stimulant. While we had every right to be near exhaustion after our thirteen trying hours this first day in 'Hurricane George,' we did not get to bed early that night. There was too much to tell, and too much to discuss concerning the flight scheduled to leave early the next morning. This second flight promised to be even more lucrative of results than the first, for we were scheduled to fly in the newest plane in the squadron. It had only 100 hours or so in the air and contained many new features the other planes didn't have. Moreover it had bomb bay tanks and could leave the ground with nearly eight thousand five hundred gallons of gasoline.

"There were a few changes in the crew but Eastburn was the pilot again on the second flight. The takeoff was scheduled for 6:30 A.M. The storm was in a critical position as far as warnings were concerned, and the Miami office was anxious to get information as early as possible upon which to base a warning for the East Coast. 'George' was located over the eastern Bahamas and was moving slowly westward, a distinct threat to the entire Eastern Seaboard but immediately to the Florida coast."

The first hint of what was in store for the hurricane hunters that day turned up as they completed their briefing at the ship and prepared to board the plane. The engineer, in a last-minute checkup, found a hydraulic leak and there was a delay of a little more than an hour before that could be repaired. Finally they pulled away from the line and out to the end of the runway. Number 4 engine was too hot. There was another delay while further checks were made into the power plant. Finally they were off—all one hundred thirty-five thousand pounds. This was to have been a very long flight and every available bit of gasoline storage had been utilized.

The plan on this day was once again to make a try for data near the top of the storm, to verify and expand the startling information gained the preceding day. This plane had de-icer boots and they were not concerned about the rime ice that might tend to accumulate, as it had the day before. First, they were anxious to get certain data from a low-level flight, and to learn how effectively the radar could be used for navigating a large plane like the B-29 near the center of the storm. They went out at ten thousand feet again but continued to a point about eighty miles north of the storm at this elevation. By this time they had crossed about four of the spiral rain bands (the spiraling arms of the "octopus"). Here the plane turned downwind parallel to another of the rain bands and flew through the corridor to within viewing distance of the eye. They gradually descended as the base of the middle-level clouds lowered near the storm center. Leveling off at seven thousand five hundred feet, they were in and out of clouds with horizontal visibility low much of the time. However, there was scarcely a thirty-second period when the crew were unable to see the sea surface below. Navigation at this stage was entirely by radar. Again the amazing thing was the lack of turbulence throughout this flight. This was a really big storm. They were flying at only seven thousand five hundred feet through one of the most violent sectors, only twenty to thirty miles from the eye itself, yet they encountered nothing that could be described as important as moderate turbulence. Simpson's early experience in hurricane flying in 1945 in a C-47 had been repeated. They were flying in comfort under conditions which gave them a command of all the information needed to report the position and intensity of the storm. Simpson remarked: "What a difference this is from the battering flights at five hundred feet in the B-17's which have been standard operating procedure ('SOP') with the squadron until this season!"

The fascination of flying in comfort so near the storm center tempted them to continue this exploration of reconnaissance tactics somewhat longer. However, there were many other important things to be done on this flight and there was no time to waste. They picked their way across one of the bands to an outer "corridor" and retreated to a point about 150 miles from the center and once again began to climb. Perhaps in the fascination of traveling so close to the eye in such comfort they had become complacent. In any case, the events which followed in fast succession left no room for further complacency. They had climbed no higher than twelve thousand feet when someone spoke on the interphone with a bit of a quiver in his voice, "I smell gasoline." The hatches were opened and the plane vented hurriedly. Eastburn went aft to investigate and returned with a worried look on his face. He spoke to the engineer, who scrambled through the tube (connecting the fore and the aft sections of the plane) on the double. It was not until after he returned, about twenty minutes later, that the rest of the crew learned that they had developed a very serious gasoline leak in one of the hoses connecting the bomb bay tanks. Nearly a thousand gallons of gasoline had been streamed through the bomb bay doors. The engineer had completed the repair satisfactorily and, after a brief consultation with the plane commander, the crew consented to go ahead with the project.

"We climbed to twenty thousand feet," said Simpson in his report. "I was seated on the jump-seat between the radar operator and the engineer, looking through the tube. I saw from the tube a wisp of smoke drifting lazily toward the aft section. I do not recall my exact reaction but I am sure I was not a picture of composure when I called this to the engineer's attention. Nor did he stop to check with the plane commander before demonstrating that he also was a handy man with a fire extinguisher. The cause was a simple thing. As we climbed, the engineer had turned on the cabin heater, the insulation of which was a bit too thin in the tube so that the padding in the tube began to smolder. Perhaps this wasn't a very important item but it didn't contribute to the peace of mind of any of the crew, especially when it was remembered that only a few minutes earlier the bomb bay gas tank immediately beneath that tube had been leaking like a sieve. Again the plane commander checked with the crew. Again, but with noticeable hesitation, it was agreed that we would proceed with the project. Higher and higher we climbed. This time we reached the forty thousand feet mark with the base of the high cirrostratus still above us. So we leveled out, trimmed our tabs and set our course for the storm center. This time we were determined to descend from forty thousand feet in the eye to get a sounding there and then return home at low levels.

"We soon reached the base of the cirrostratus and entered the clouds. The de-icers were working. Again the data began to roll in along the same pattern as observed the previous day—at least for several minutes, until the interphone was filled with the excited voice of the right scanner with a spine-tingling report to the commander, 'Black smoke and flame coming from number 4.' At the same time the plane began to throb, roll and yaw. In less time than it takes to say it, the 'boys' in the front compartment of this B-29 became *mature men*—wise, efficient, stout-hearted men, each with a job to do and each one doing it with calculated deliberateness, yet speedily. There was grim determination here but no evidence of emotion. This magnificent tribute to topnotch training had an exhilarating effect upon me and tempered to some extent the abashment which I could not help feeling as a result of my helplessness in this situation, and the fear which clutched my heart.

"We were lucky! The single carbon dioxide charge released by the engineer extinguished the fire in the engine. Number 4 was feathered and began to cool but our troubles were far from over. The engineer had manuals and technical orders spread out on all sides of him and was working feverishly to restore some power to number 4, as the indicated air speed dwindled from 168 to 166 to 164 or 5, hovering precariously above the deadly stallout at 163. We were only a few miles from north of the center by this time but no one had recorded the data. We were too busy worrying. The pilot was in the process of putting the plane into a long glide to increase the air speed, when the left scanner claimed the interphone circuit with, 'Black smoke and flame coming from number 1.' This time we *were* in real trouble. However, the engineer had anticipated further difficulty and was ready again. It was only a matter of seconds before the fire was out and some semblance of power had been returned to number 1. But we were still five hundred miles from the nearest land and very near the center of a granddaddy of hurricanes. So we declared an emergency and headed for MacDill Field."

Altogether, this was an ironical turn of affairs. An old plane had acted like a lady the day before and now a new one had frightened the crew with its mechanical troubles, but the newer methods of hurricane hunting, the "tricks of the trade," had fortunately taken some of the danger out of the storm itself. Otherwise the mechanical troubles might have combined with the weather to spell disaster.

12. TRAILING THE TERRIBLE TYPHOON

>>

"The workshop of Nature in her wildest mood."—Deppermann

So far as anyone knows, the most furious of the typhoons of the Pacific are no bigger or more violent than the worst of the huge hurricanes of the Atlantic and the West Indies. They belong to the same death-dealing breed of storms, but the typhoons come from the bigger ocean; they sweep majestically across these vast waters toward the world's largest continent; and to the south and southeast lies a longer stretch of hot tropical seas than anywhere else on earth. Perhaps it is the enormous extent of the environment that explains the fact that in the average year there are three or four times as many Pacific typhoons as there are West Indian hurricanes. The greater excess of energy generated in this enormous Pacific storm region by hot sun on slow-moving waters is evidently released by a more frequent rather than a more violent dissolution of the stability of the atmosphere.

But there is something about typhoons that causes the people to look upon them with even greater terror than in the case of hurricanes. Likewise, the storm hunters tackle the job of tracking them with less confidence. Typhoons come from greater distances. Their points of origin may be scattered over a wider area. Much more often than is the case with hurricanes, there may be two or more at the same time. In their paths of devastation they fan out over a bigger and more populous part of the world. It takes more planes, more men and longer flights to keep up with typhoons than with hurricanes.

For many decades the people of the Far East struggled valiantly against the typhoon menace without much interest on the part of the Western World. Native observers reported them when they showed their first dangerous signs and then came roaring by the islands in the Pacific, including the Philippines, as they swept a path of devastation on the way to China or Japan. Men on ships equipped with radio sent frantic weather messages to Manila, Shanghai or Tokyo as they were being battered by monstrous winds and seas. Father Charles Deppermann, S.J., formerly of the Philippine Weather Bureau, who did as much as any man to help people prepare for these catastrophes, made an investigation to see why some of the typhoon reports from native observers were defective. He listed a few of the reasons.

One observer said his house was shaking so much in the storm that he was unable to finish the observation. He added that ninety per cent of the houses around him were thrown to the ground. Another common complaint was that the observers could not read the thermometers because the air was full of flying tin and wood. Another apologetic man put on the end of his observation a note that the roof of the weather station was off and the sea was coming in. The observer on the Island of Yap fled to the Catholic rectory and looked back to see his roof, walls, and doors blowing away, but he sent his record to the forecast office! Another observer on Yap was reading the barometer when it was hit by a flying piece of wood and the observer was knocked to the floor. One of the observers had excuses for a poor observation because he had to run against the wind in water knee deep. In another place, the wind blew two rooms off the observer's house at observation time. But the most convincing excuse for failure was from another town where the observer was drowned in a typhoon before the record was finished.

It is a strange fact, too, that one can look at all these records and the reports written by the Pacific storm hunters after they got going, and seldom see a vivid description of the fearful conditions in the typhoon. The white clouds turning grayish and then copper-colored or red at sunset. The rain squalls carried furiously along. The roar of giant winds and the booming sea as the typhoon takes possession of its empire in huge spirals of destruction. With death and ruin on all sides, nobody seemed to have the energy to write about it. The tumult passed, the wind subsided, the water went out slowly, and the observer wrote a brief apology for the bedraggled condition of the records.

In the same way, the typhoon hunters let their planes down at home base too tired to do anything except compile a few technical notes. The vastness of the thing seemed to leave them speechless. The plane went out on a mission and the base soon vanished, a shrinking dot on the horizon. The mind tired of thinking about the near-infinite expanse of Pacific waters, of thinking about running out of fuel in an endless search of winds, clouds and waves, of thinking about never getting back to that little dot beyond the horizon.

Into this ominous arena the American fleet nosed its way, island by island, in the war against the Japanese. By methods which had been handed down from older generations, strengthened by all the modern improvements that could be added, the Americans tried to keep track of tropical storms in this enormous region where trade winds, monsoons and tropical winds hold their several courses across seemingly endless seas, but here and there run into conflict or converge in chaos. Twice when their predictions were not very good, the fleet suffered and in the second instance the typhoon humbled the greatest fleet that ever was assembled on the high seas. The Commander-in-Chief, Pacific, demanded reconnaissance without delay. As men do in time of war, the Navy aerologists moved swiftly and effectively to meet the challenge. In fact, they had anticipated it in part and had plans in the blue-print stage, even before the big Third Fleet took its brutal beating in December, 1944.

Most of the stimulus came from the Atlantic side, where organized hurricane hunting had begun in the middle of the year. But it was not long until the Japanese were driven out of the typhoon areas. In June, 1945, they were being blasted out of Okinawa as typhoon reconnaissance was beginning. In fact, the first men to go out to penetrate a typhoon had to be careful to keep away from Okinawa. By that time the Japanese had committed all their fading sea and air power, including their last remaining battleship, to the defense of Okinawa, and after June, the U. S. Navy had no real enemy except the typhoon.

Beginning in June, 1945, the Navy airmen and aerologists flew two kinds of missions. Almost daily they went out to check the weather, and if they found a full-grown typhoon or one in formation in an advanced stage, special reccos were sent out. One flight went out as soon as it was daylight and the second took off about six hours afterward, early enough to make sure that the second would be completed by nightfall. This was rather tough going. As one of the aerologists pointed out, Pacific distances were so large that if they were considered in terms of similar distances in the United States, a common mission would be like a take-off from Memphis and a search of the area of a triangle extending from Washington, D. C., to New York City and back to Memphis.

Aircraft used by the Navy were Catalinas (PBY's), Liberators (PB4Y-1's), and Privateers (PB4Y-2's). All were four-engined, land-based bombers, some fitted with extra gasoline tanks for long ranges. Before leaving base in the Philippines or the Marianas, the aerologists briefed the crews. In flight, the aerologist directed changes in the course of the plane, but the pilot could use his own judgment at any time when he thought the change might exceed operational safety. From June through September, 1945, the Navy flew a total of one hundred typhoon missions, averaging ten hours each. Lieutenants Paul A. Humphrey (a Weather Bureau scientist after the war) and Robert C. Fite, both of whom flew constantly on these missions, gathered data from all flight crews, and at the end of the season wrote descriptions of five typhoons which were more or less typical.

Some of the most interesting of these missions were directed into the big typhoon which came from the east, crossed Luzon in the Philippines and roared into the China Sea, in the early part of August. On the fourth of the month, one of the Catalinas was checking the weather three hundred miles east of Leyte and saw a low pressure system developing a small tropical disturbance. It grew, was checked daily, and on the sixth blew across Luzon and reached its greatest fury in the South China Sea on the seventh.

The first plane that went into the typhoon in this position was directed to the right and north of the center, to take advantage of tail winds and to spiral gradually into the center. As it approached the center, the plane climbed to about five thousand feet, and the crew had a beautiful panoramic view of the clouds piled up on the outer rim of the eye. On account of the awful severity of the turbulence the plane had experienced around the eye, they descended again and flew to home base at altitudes between two hundred and three hundred feet.

On examination of the aircraft after the battered crew had let down at home base, it was found that the control cables were permanently loosened, the skin on the bottom of the port elevator fin had been cracked away from the fuselage, one Plexiglas window was bowed inward, and the paint was removed from the leading edges. Because of the violence of turbulence on this flight, the nervous crew of the second recco plane on that day was instructed to reconnoiter but not to try to go into the center.

On the fifth of September a violent typhoon formed between the Philippines and Palau and moved northwestward toward Formosa. On the tenth a recco plane ran into trouble in this storm. Twice while flying at two thousand feet, it met severe downdrafts, losing altitude at five hundred to one thousand feet per minute while nosed upward and climbing at full power. The eddy turbulence was extremely severe and most of the crew members became sick. The second recco plane on that date ran into violent turbulence also, and at times it was almost impossible for the pilot and co-pilot to keep the plane under control.

And then disaster struck! By the end of September the Navy storm hunters had gone out on one hundred missions into the hearts of typhoons and, although many of them had been frightened and badly battered, there had been no casualties. They made up a report as of September 30, commenting on their phenomenal good fortune on these many flights. But on the very next day, October 1, one of the crews which had been making these perilous missions departed on a flight into a typhoon over the China Sea. Those men never came back. No one had any idea as to what had actually happened, but the members of other crews could well imagine what might have happened, and whatever it was, it must have ended in typhoon swept waters where none of the storm hunters expected to have any chance of survival. It could have happened in the powerful winds around the eye or in one of those bands extending spirally outward from the center, filled with tremendous squalls and fraught with danger to brave men venturing into these monstrous cyclones of the Pacific. The report—even before this sequel—had stressed the hazardous nature of reconnaissance.

In these Pacific missions, the pilots and aerologists, even without radar, had become aware of the doughnut-shaped body of the storm with squall bands spiraling outward (the octopus arms). But they got very little information that they thought would help in predicting the movements of typhoons, except the old rule that the storm is likely to continue on its course unchanged, tending to follow the average path for the season. The explorations by aircraft as a means of getting data were far more useful in locating storms and determining their tracks, however, than any other methods.

After the end of 1945, the reconnaissance of tropical storms, both in the Atlantic and the Pacific, was in trouble, owing to demobilization. Many experienced men returned to civil life and it was necessary to start training all over again. The Navy set up schools for two squadrons of Pacific storm hunters late in 1945, at Camp Kearney in California. The graduates were in action in 1946.

After the surrender of the Japanese, the Air Corps maintained a Weather Wing in the Pacific, with headquarters in Tokyo. Part of its job was to give warnings of typhoons threatening Okinawa, where the United States had established a big military base. Here they thought they had built structures strong enough to withstand typhoons, but they learned some bitter lessons. The most violent of all the typhoons of this period was one named "Gloria" which almost wiped Okinawa clean in July, 1949.

A most unusual incident occurred over the Island of Okinawa when the center of Gloria was passing. The Air Force was short of planes in safe condition for recco, but managed to get enough data to indicate the force and probable arrival of this violent typhoon. It happened that Captain Roy Ladd, commander of Flight #3, was in the area, with Colonel Thomas Moorman on board, making an inspection of recco procedures in the area. Their report gave the following information:

"As Gloria roared over a helpless and prostrate Okinawa, weather reconnaissance members of Crew B-1 circled in the eye of the big blow and watched the destruction of the island while talking to another eyewitness on the ground. That hapless human was the duty operator for Okinawa Flight Control, who, despite the fact that his world was literally disappearing before his eyes and the roof ripping off overhead, nevertheless stuck to his post and eventually contacted three aircraft flying within the control zone and cleared them to other bases away from the storm's path."

Describing the situation, Captain Ladd stated that he had attempted radio contact with Okinawa for some time but was prevented from doing so by severe atmospheric conditions. After a connection had been established, one hundred miles out from Okinawa's east coast, the control operator requested them to contact two other aircraft in the area and advise them to communicate with Tokyo Control for further instructions.

Shortly thereafter, the RB-29 broke through heavy cloud

formations into the comparatively clear eye of the big typhoon. The southern tip of the island became visible, just under the western edge of Gloria's core. Gigantic swells were breaking upon the coast and the control operator advised that winds had been 105 miles per hour just thirty minutes before and had been increasing rapidly. He reported that the control building's roof had just blown off, all types of debris were flying by, and aircraft were being tossed about like toys.

A little later, the ground operator had to crawl under a table to get shelter because nearly all of the building had been blown away, bit by bit. Structures of the quonset type were crushed like matchboxes and carried away like pieces of paper. Their roofs were ripped like rags. A cook at the Air Force Base hurried into a large walk-in refrigerator when everything began to blow away. "It was the only safe place I could find," he explained afterward. "The building blew away but the refrigerator was left behind and here I am."

One of the meanest of the typhoons of this period was known as "Vulture Charlie." It was dangerous to airmen because of the extreme violence of its turbulence. Ordinarily, the typhoons were known by girls' names, and for that reason the typhoon hunters in the Pacific were known as "girl-chasers." But "Vulture Charlie" got the first word of its name from the type of mission involved, and "Charlie" from the third word in the phonetic alphabet used in communications.

On November 4, 1948, an aircraft commanded by Captain Louis J. Desandro ran into the violent turmoil of Vulture Charlie and described it as follows:

"We hit heavy rain and suddenly the airspeed and rate of climb began to increase alarmingly and reached a maximum of 260 miles per hour and four thousand feet per minute climb to an altitude of three thousand seven hundred feet. The sudden increase in altitude was brought about by disengaging the elevator control of the auto-pilot and raising the nose to control the airspeed. Power was not reduced because of our low altitude. After about thirty seconds to one minute of this unusual condition we hit a terrific bump which appeared to be the result of breaking out of a thunderhead. The airspeed then decreased to 130 miles per hour in a few seconds due to the fact that we encountered downdrafts on the outer portion of the thunderhead and were momentarily suspended in air. At this point the left wing dropped slightly and I immediately shoved the nose down to regain airspeed. Before a safe airspeed was again reached, we had descended to an altitude of one thousand one hundred feet.

"As a result of this turbulence my feet came up off the rudder pedals. The engineer, who was sitting on the nose wheel door instructing a student engineer, came up off the floor like he was floating in the air. The navigator and weather observer were raised out of their seats. A coffee cup, which was on the back of the airplane commander's instrument panel, was raised to the ceiling and came down on the weather observer's table. Cabin airflow was being used and the airflow meter exploded and glass hit both engineers in the face."

In December, 1948, a crew under the command of Lieutenant David Lykins was instructed to use the boxing procedure in a typhoon called "Beverly." On one of their missions, they flew into it on December 7. The following is based on his report:

The operations office instructed the crew to climb to the seven hundred millibar level (about ten thousand feet) after take-off, penetrate the eye of the storm, take a fix in the center, then make a spiral descent and sounding down to one thousand five hundred feet and proceed out of the storm on a northwesterly heading, to begin the pattern around the storm center.

After the briefing, the crew ate dinner, while talking anxiously about the trip, and returned to the aircraft to load personal equipment. When they were airborne with the gear and flaps up, they made an initial contact with Guam Control. There was no reported traffic, so they were cleared. The instructions were complied with and a heading of 270 degrees was taken up. Soon there was discernible on the horizon a vast coverage of high, thin clouds at about thirty thousand feet. This indicated the presence of the storm, verified by the south wind and slight swells that were perpendicular to the flight direction of the plane. The wind was increasing and the swells were noticed to intensify. The boundary of the storm area was very distinct as they approached the edge. At this point, the surface wind was estimated to be thirty-five knots from 180 degrees.

A few minutes later they were on one hundred per cent instrument flying conditions and the moderate to heavy rain and moderate turbulence persisted until they missed the eye and flew south for fifteen minutes. Because they were on instruments and could not see the surface, they were unable to determine the highest wind velocity in the storm. It was estimated close to one hundred knots. At this point they noticed that they had a good drift correction for hitting the center satisfactorily, so they held the 270 degrees heading, relying on the radar observer to be able to see the eye on the scope.

Approximately fifteen or twenty minutes later, the radar observer reported seeing a semi-circular ring of clouds about twenty-five degrees to the right at about twenty-five miles range. The same kind of ring was detected to the left, about the same distance, however. Figuring they had drifted to the right of the center, they elected to intercept the left center seen on the radar and flew until they received an ill-omened pressure rise, when it was apparent they had made a wrong choice!

To make sure they were not chasing circular rings of heavy clouds or false eyes on the scope, they made a turn to 180 degrees and held it long enough to enable them to see the surface wind. After about ten minutes they saw the surface and judged the wind to be coming from approximately west-northwest. They headed back for the center of the storm with the wind off their left wing, allowing fifteen to twenty degrees for drift. In approximately fifteen minutes the radar observer reported the eye as being almost directly ahead. Lieutenant Lykens said:

"At 0906Z (1906 Guam time) we broke out into the most beautiful and well-defined eye that I have ever seen. It was a perfect circle about thirty miles in diameter and beautifully clear overhead. The sides sloped gently inward toward the bottom from twenty-five thousand feet and appeared to be formed by a solid cloud layer down to approximately five thousand feet. From one thousand feet to five thousand feet were tiers of circular cumulus clouds giving the effect of seats in a huge stadium."

They descended in the eye, made their observations and then prepared to depart. Lieutenant Lykins continued:

"As we entered the edge of the eye we were shaken by turbulence so severe that it took both pilots to keep the airplane in an upright attitude. At times the updrafts and downdrafts were so severe that I was forced down in my seat so hard that I could not lift my head and I could not see the instruments. Other times I was thrown against my safety belt so hard that my arms and legs were of no use momentarily, and I was unable to exert pressure on the controls. All I could do was use the artificial horizon momentarily until I could see and interpret the rest of the instruments. These violent forces were not of long duration fortunately, for had they been it would have been physically impossible to control the airplane.

"Since the updrafts and downdrafts were so severe, we were unable to maintain control of the altitude; all we could do was to hold the airspeed within limits to keep the airplane from tearing up from too much speed or from stalling out from too little. After the first few seconds, we managed to have the third pilot, who was riding on the flight deck, advance the RPM to 2400 so we could use extra power in the downdrafts, and so we could start a gradual ascent from the area. Neither of us at the controls dared leave them long enough to do it ourselves.

"The third pilot received a lump on his forehead when he struck the rear of the pilot's seat, and bruised his shoulder from another source in doing so. Since he had no safety belt, he was thrown all over the flight deck.

"This area of severe turbulence lasted between five and six minutes and every second during this time it was all both of us could do to keep the airplane in a safe attitude and to keep it within safe airspeed limits and maintain a general heading.

"It is almost impossible for me to describe accurately or to exaggerate the severity of the turbulence we encountered. To some it may sound exaggerated and utterly fantastic, but to me it was a fight for life.

"I have flown many weather missions in my thirty months in the 514th Reconnaissance Squadron, I have flown night combat missions in rough winter weather out of England, and I have instructed instrument flying in the States, but never have I even dreamed of such turbulence as we encountered in typhoon Beverly. It is amazing to me that our ship held together as it did."

When the severest turbulence subsided the hurricane hunters found they had gained an altitude of about six thousand feet. At this point they decided to climb to 10,500 feet and proceed directly to Clark Field. It was night time and, since they were shaken up pretty badly, this seemed the most sensible course of action to be taken. They had no way of knowing the extent of any damage they might have sustained. The engineer reported that the booster pumps had all gone into high boost; one generator had quit. The radar observer said that the rear of the airplane was a mass of rubble from upturned floorboards, personal equipment, sustenance kits, and such. The flight deck had extra equipment all over it. In addition, the co-pilot had twisted off a fluorescent light rheostat switch when the plane hit the turbulence as he was adjusting it. The radar observer reported his camera had been knocked to the floor.

After his experience in leaving the eye of Beverly at one thousand five hundred feet, the lieutenant had one statement to make and he said it could not be overemphasized.

"An airplane with human beings aboard should never be required to fly through the eye of a typhoon at an altitude below ten thousand feet. If a pattern must be flown at one thousand five hundred feet in the storm area, it should be clearly indicated that the area of the eye be left at the seven hundred millibar level and the descent be made at a distance of not less than seventy miles from the center. Full use of radar equipment should be exercised in avoiding any doubtful areas."

On inspection after landing, the following damage to the airplane was found: A bent vertical fin, warped flaps, tears in fairing joining the wing and fuselage, untold snapped rivets on all parts of the airplane, fuselage apparently twisted, and one unit in the center of the bomb bay was torn from its mountings.

Reports of this kind leave some doubt as to whether the typhoon actually is not more violent than the West Indian hurricane.

Another typhoon of extraordinary violence which gave the storm hunters serious trouble struck Wake Island on September 16, 1952. Wake is a little island in the Pacific Ocean, a small dot on the map, the only stopping-place between the Hawaiian Islands, more than two thousand miles to the eastward, and the Marianas, more than one thousand miles to the westward. This spot, a stop for Pan American planes, was captured by the Japanese and then recaptured by the United States in World War II. When the Korean War opened, military planes used this small island as a refueling place en route from the Pacific Coast of the United States to Japan.

Before taking off from Honolulu, the airmen wanted a forecast for this long route and a report of the weather at Wake. Also, before taking off from Wake, they asked for a forecast for the trip to the next stop at Guam, Manila or Tokyo. The military called on the Weather Bureau and Civil Aeronautics Administration to furnish the weather service and the communications. They started operations at Wake very soon. By 1952 men from these two agencies were on the island, some with their wives and children. The Standard Oil Company and Pan American Airways also had people there. For the most part, they were housed in quonset-type structures, but some old pillboxes constructed during the war still dotted the island and could be used for refuge from typhoons if the wind-driven seas did not rise high enough to flood them. There were only three concrete buildings and they were used for offices and storage.

On the morning of September 11, 1952, the Weather Bureau forecaster drew a low center on his weather chart far to the southeast of Wake. His analysis was based largely on two isolated ship reports, the only information available from a one million square-mile ocean area lying to the east-southeast of his tiny island station. Here was just enough data to arouse suspicion and alarm that a developing tropical disturbance was somewhere—anywhere—within this vast expanse of sea and air; but not enough information to indicate a position, or probable intensity, or actually to confirm the existence of a well-defined storm.

During the next three days, the question of continuing the low on successive charts, and the problem of deciding its position, were mostly matters of guesswork on the part of the Weather Bureau staff at Wake; there was only one ship report from the critical area during the time. Then on September 14 the existence of a vortex was established. A single ship report, together with reports from Kwajalein and Eniwetok, gave good evidence of cyclonic circulation.

From this time on, until the storm struck at daybreak on the sixteenth, everybody on the island worried about it, and the weathermen went all out in tracking it and disseminating information. Meanwhile the typhoon—which had been named "Olive"—grew into the most destructive storm to hit Wake since it was first inhabited in 1935. The forecasters' job was a difficult one because of meager observational data. There were heartbreaking delays in securing airplane reconnaissance due to mechanical breakdown that grounded the B-29 stationed at Wake for that purpose until an engine part could be flown in from Tokyo.

Early on the morning of the sixteenth, strong winds of the typhoon began to sweep across the island, a very rough sea was breaking on the shores, and debris was flying through the air. One can easily imagine the alarm of these people in the vast Pacific, on a tiny island beginning to shrink as the waters rose, and giving up its soil, rocks, and parts of buildings to the furious winds, steadily increasing.

A large power line fell across several quonsets just north of the terminal building, and huge sparks began flying where they touched the Weather Bureau warehouse.

The account which follows is condensed from the report made by the Weather Bureau man in charge, Walton Follansbee:

The wind indicators in the Weather Station shorted out early, and expensive radiosonde and solar radiation equipment was badly burned by the runaway power. The indicators in the tower, however, remained operative until the last weatherman abandoned it. They took turns climbing the tower steps to check the velocities, calling the readings off over the interphone from tower to weather station. On Follansbee's last trip to the tower, the strongest gusts observed were eighty-two miles per hour, although one of the observers had caught gusts to ninety miles per hour shortly before. The strain on the structure was sereve, and he was happy to get down the stairs safely. Soon afterward, Jim Champion, observational supervisor, took full responsibility for this unwanted task. He then reported over the interphone that the wind was north-northwest at eighty miles per hour with gusts to 110. Follansbee advised him to abandon the tower. He replied that he believed he was safer staying there than trying to come down the stairs, which were wide open to the elements. He was told to use his own judgment, since it was his life at stake.

Women and children had been taken to the terminal building or other safer places than the quonsets, which now began to break up. Anybody who ventured in the open was likely to be blown off his feet and that was exceedingly dangerous, for the sea was close by, and now and then the roof of a quonset went off and was carried dangerously across the island and out to sea. Winds of hurricane force blew the water from the lagoon which began engulfing the south and east parts of the island. The wind reached a steady velocity of 120 miles an hour, with gusts up to 142 at the height of the storm.

By that time, most of the women and children were huddled in the operations building and they were terrified when the roof went off, leaving them exposed to the torrential rain and furious winds, but the walls held. About this time, a report was received from a reconnaissance plane that had come from Guam and made its way into the center of the typhoon. The crew put the center about thirty-five miles northeast of Wake but said the plane was suffering structural damage and was heading for Kwajalein.

By evening the winds were subsiding and a check showed that owing to such preparations as they had been able to make and the constant struggle of all on the island to prevent disaster, not a single life was lost and no one was seriously injured. Wake Island, however, was a shambles, and there was very little food not contaminated and practically no drinking water. The water distillation plant had been destroyed.

But soon one of the Air Force B-29 planes ordinarily used in typhoon reconnaissance flew in from Kwajalein and brought three hundred gallons of water in GI cans lashed to the bomb bays and two tons of rations for distribution to the battered and hungry people of Wake Island. Before long, the little island was back in business, serving the big planes on the way from Hawaii to the Far East.

13. GUEST ON A HAIRY HOP

"On the rushing of the wings of the wind. It is indeed a knowledge which must be felt to be in its very essence full of the soul of the beautiful."—Ruskin

A hurricane flight which proves to be rougher than usual is known among the hunters as a "hairy hop." It is an amazing fact that there are men who want to come down to the airfield when a big storm is imminent and "thumb a ride." Mostly, they are newspaper reporters, magazine writers, photographers, civilian weathermen, and radio and television people. Usually they are accommodated, if they have made arrangements in advance. Some of these rides have been quiet, like a sightseer's trip over a city, while others have been "hairy."

One of the first newspapermen to take a ride into a full-fledged hurricane was Milt Sosin of the Miami *Daily News.* In 1944, Milt read about men of the Army and Navy who were just beginning to fly into hurricanes and he became obsessed with the wish to go along. When he asked for permission, the editor said "No" in a very positive tone. He could see no point in having a good staff correspondent dropped in the ocean during a wild ride in a hurricane. Sosin insisted and he was told to see the managing editor. He did and there was another argument. Sosin told him, "If I don't, somebody else will and we'll be scooped." Reluctantly, the managing editor gave permission. But when Sosin asked the immigration authorities, they said "No. You have no passport, and you don't know what country you may fall in." They refused. Sosin hung around and argued. He pointed out that if the plane went down at sea, he wouldn't need any passport to the place he was going, and they finally agreed.

Milt Sosin got his wish in full measure on September 13, 1944, in the Great Atlantic Hurricane which had developed a fury seldom attained, even in the worst of these tropical giants. It had crossed the northern Bahamas and was headed northwestward on a broad arc that was to bring its death-dealing winds to New Jersey, Long Island and New England. Already we have told the story of Army and Navy planes probing this big storm, including the pioneering trip by Colonel Wood and others of the Washington weather staff. At the end of this trip, Sosin was glad to be back on land and vowed, "Never again!" But, somehow, he still had the urge to see these storms from the inside and afterward was a frequent guest of the Navy and Air Force.

One of Sosin's most interesting trips was on September 14, 1947, in a B-17. They took off from Miami. Al Topel, also from the Miami *Daily News,* went along to take pictures, and Fred Clampitt, news editor of Radio Station WIOD, was the other guest. The big hurricane was roaring toward the Bahamas with steadily increasing fury and the people of Florida were worried—and for good reason, for three days later it raked the state from east to west, killing more than fifty people and causing destruction estimated in excess of one hundred million dollars. By many observers it was eventually rated as the most violent hurricane between 1944 and 1949.

They ran into it east of the Bahamas. As the plane burrowed its way through the seething blasts, Sosin wrote in his shaking notebook:

"This airplane feels as if it's cracking up. Ominous crashes in the aft compartment accompany every sickening lurch and dive as, buffeted by 140-mile-an-hour winds and sucked into powerful downdrafts, the huge bomber bores through to the core of the storm."

Sosin said that the pilot, Captain Vince Huegele, and the co-pilot, Lieutenant Don Ketcham, were literally wrestling with the hurricane in clothes sopping wet from perspiration and, as soon as they came into the center, began to take off their wet garments. Ketcham had "pealed down to his shorts before the plane plunged back into the mad vortex."

At this point they were surprised to see another plane in the storm, a B-29, flying in the eye at thirty-six thousand feet, trying to discover the "steering level" where the main currents of the atmosphere control the forward movement of tropical disturbances such as this one. The radio man, Sergeant Jeff Thornton, was trying to contact the B-29, miles overhead, but with no luck. Sosin wrote in his notebook:

"But here at this low level we have more to worry about than trying to reach the other plane. We are getting an awful kicking around. Wow! That was a beaut. Al Topel was foolish enough to unfasten his safety belt and stand up for a better angle shot of the raging turbulent sea below. We must have dropped one hundred feet and his head hit the aluminum ribbing of the plane's ceiling. Then, trying to protect his camera, he skinned his elbows and knuckles. Now he's given up and has even strapped a safety belt around his camera."

The crew was busy plotting positions and checking on the

engines. To them it was an old story, except that none could recall such violent turbulence. The craft was low enough for them to get glimpses of the sea but they wanted a better view and they began to descend cautiously. Sosin wrote:

"The turbulence is getting worse. The sea is streaked with greenish-gray lines which look like daubs made by a child who has stuck his fingers into a can of paint. Now we are closed in. We are flying blind. Capt. John C. Mays, the weather observer, starts giving the pilots readings from his radar altimeter while Huegele sends the plane lower and lower in an effort to establish visual contact with the sea.

" 'Five hundred feet,' Mays calls into the plane's intercom.

" 'OK,' replies the skipper.

" 'Four hundred feet.'

" 'Roger.'

" 'Three-fifty.'

" 'Roger.'

" 'Two-fifty.'

" 'OK.'

" 'Two hundred feet,' Mays' voice is still even.

" 'OK,' comes Huegele's voice.

"It may be OK with him but it isn't with me. I just found myself tugging tentatively on the pull toggles which will inflate my 'Mae West' life jacket if I yank hard enough. I checked a long time ago to make certain the CO cartridges were where they should be.

"Fred Clampitt, WIOD news editor, is turning green.

"No, it's not fear. He's sweating so much that the colored chemical shark repellent in a pocket of his life jacket is starting to run.

"Then we sight the sea again. From this low level the waves are frightening. They are traveling in all directions, not in just one, and they break against each other, dashing salt spray high into the air. It's all too close.

"Now the ceiling is lifting and we are climbing—250, 300, 500, 700 and we level off. It grows less turbulent and Observer Mays looks up from his deep concentration.

" 'I may be wrong,' he says, 'but it looks to me as if it's made a little curve toward the north.'

"Which is very interesting—but more interesting is the fact that the day's work is over and we're on our way home."

In 1947, the Air Forces were assigning B-29's to their Kindley Base at Bermuda, to replace the B-17's. The big superforts had room for guests and it soon became common to have somebody hanging around Kindley to get a ride. When a big storm was spotted east of the Windward Islands on the eleventh of September of that year, two newspaper reporters and a photographer from *Life* Magazine, Francis Miller, were waiting at Bermuda for a hop. The big hurricane became even more violent as it turned toward the southwest and swept across Florida. It was September 14th when Milt Sosin of the Miami *Daily News* got his "hairy hop" in this same blow. As it crossed the coast, winds of full hurricane force stretched over a distance of 240 miles and the wind reached 155 miles an hour at Hillsboro Light. By this time the hurricane hunters were fully occupied and the riders were left on the ground. Miami communication lines were wiped out and control of the hunters had been shifted to Washington. In charge of a B-17 at Bermuda was Major Hawley. His co-pilot was Captain Dunn, who had learned hurricane hunting in "Kappler's Hurricane" and other earlier storms. Late on the seventeenth, as the storm roared across Florida with night closing in, Hawley had heard nothing from Washington about his plane going into it, so he gave up and told the riders to come back in the morning.

Early the next morning, one of the reporters, a staff writer for the Bermuda *Royal Gazette*, was sitting around in his shorts and thinking about breakfast when Lieutenant Cronin rushed in and said they were ready to take off. The reporter started to get dressed, but Cronin said, "Let's go. Just as you are. You may drown but you won't freeze." They stopped in Hamilton, got the other reporter and the photographer, and found Hawley walking up and down, impatiently waiting for last instructions. So the reporter took a trip of 3,350 miles in his shorts and had a bird's-eye view of the southern Seaboard, the Atlantic Ocean and Gulf of Mexico and a bad-acting hurricane.

It was a "hairy hop." They had orders to refuel at Mobile, so they put down at the airfield there, all other planes having been evacuated the day before. An Air Force man came out and asked, "Where you goin'?" They told him and he turned around and shouted, "Some dang fools think they have a kite and can fly through a hurricane." More men came out and they got gas in the plane. One big fellow said, "You can have your dern trip. But keep the storm away from here." In twenty minutes they were in the storm. The crew members were bare to the waist, perspiration pouring down, water coming through the panel joints, and everything was wet and shaking. One of the reporters described it this way:

"Suddenly the plane keeled over on one side, the left wing tip dipped down vertically, and for a moment I thought the end had come. I gulped for breath as the plane dropped. The sea rushed up towards us; huge waves reared up and mocked us, clawing up at the wing tip as if trying to swallow us in one. A greater burst from the engines, a hovering sensation for a second and then, with the whole plane shuddering under the strain, our nose once again tilted upward. I felt weak and with difficulty breathed again."

The plane had no radar and the crew had a lot of trouble trying to locate the center of the hurricane. The forecasters at Miami were anxious for an accurate position of the center. At that time airborne radars were being installed as standard equipment as rapidly as they could get around to it but the B-17's came last. Low pressure guided them, and they were trying to get into the part of the hurricane where they found the pressure falling rapidly. It was a big storm and they were having little luck in the search. "Lashed by winds and rain, the B-17 staggered across the sky," one of the reporters said afterward. He went on to tell his story:

"I was growing sick in the bomb aimer's bay stretched over a pile of parachutes and hanging onto the navigator's chair for dear life. Some baggage, roped down beforehand, now lay strewn across the gangway. Parachutes, life jackets, water cans and camera cases were thrown about into heaps. The photographer, trying in vain to take pictures out of the window, was knocked down and sent flying across the fuselage. His arms were bruised from repeated efforts. My stomach was everywhere but where it should have been. Everything went black. The plane was thrown from side to side and the floor under my feet dropped. We emerged from a big cloud into an eerie and uncanny pink half-light. The photographer clambered from the floor and tried to look out. He thought the reddish light was an engine on fire.

"Before we touched down at Tampa, after four hours of flying around in the hurricane, we reporters and the photographer were exhausted. And even then they had failed to get into the calm center, although they had sent back to Washington a lot of useful information on the storm's position."

More than anything else, the preliminaries unnerve the guest rider. They tell him about the "ditching" procedures; that is, what to do if the plane is on the verge of settling down on the raging sea. Two or three hours before take-off they are likely to have a ditching drill, along with the briefing on the storm and the check on the equipment. The guest is told that if they bail out, he will go through a forward bomb bay door. There is hollow laughter as someone makes it clear that there is very little chance of survival. But they want the guest to have every advantage.

Commander N. Brango of Navy reconnaissance says: "Yes, we get a good many requests from men who want to go along. Would you like to go on an eight- to ten-hour flight in a four-engine, thirty-ton, Navy patrol plane? You will probably see some of the beautifully lush islands of the Antilles chain, waters shading gradually from pale green to a deep clear emerald, shining white coral beaches, native villages buried in tropical jungles, and many other sights usually referred to in the travel advertisements.

"Doesn't that sound enticing? There is just one catch. You may have to spend four to five hours of your flight-time shuddering and shaking around in the aircraft like an ice cube in a cocktail shaker, with rain driving into a hundred previously undiscovered leaks in the plane and thence down the nearest neck. You may bump your head, or other more padded portions of your anatomy, on various and sundry projecting pieces of metal (of which there seem to be at least a million). You may not be able to see much of anything, at times, since it will be raining so hard that your horizontal visibility will be nil, or you may be able to catch glimpses, straight down about 300 feet, of mountainous waves and an ocean being torn apart by winds of 90 to 150 miles per hour. There's one thing I will guarantee you, you won't be writing postcards to your friends saying, 'Having a wonderful time, wish you were here,' because you won't be able to keep the pen on paper long enough to write much of anything."

You have guessed by now that the carefully phrased invitation was just a trap to get you aboard one of the Navy's "Hurricane Hunter" patrol planes as it departs on a hurricane reconnaissance mission. According to Brango, these flights have been described by visiting correspondents, using "thrilling," "awe-inspiring," "terrifying," and other equally impressive adjectives. Actually, it is difficult to find words to describe such a flight. That it is hazardous is obvious, but the feeling that accomplishing the mission may mean the saving of many lives and much property makes it seem worth doing—not to mention the lift received from an occasional "well-done" from up the line.

Just to indicate to the prospective guest what it may be like, Brango gives "Caribbean Charlie" of 1951 as an example.

Charlie was spawned several hundred miles east of the Windward Island of Trinidad. The first notice the Navy had of its presence was a ship reporting an area of bad weather, and almost immediately one of the hurricane hunter planes from the advanced base in Puerto Rico was in the air to get the first reports on Charlie. For the next nine days Charlie led them a wild, if not a merry chase. He slipped by night through the Windward Islands and into the Caribbean, loafed across this broad expanse of water, then slammed into Kingston, Jamaica, dealing that city one of its most devastating blows in history. Then Charlie headed across the Yucatan Channel and over the Yucatan Peninsula, where he lost some of his push. Some sixteen hours later he broke into the Gulf of Campeche with renewed fury, stormed across the Gulf and into the Mexican coast at Tampico, on August 22, again costing lives and millions in property damage.

During his long rampage, he was being invaded almost daily by Navy planes. On Tuesday, August 21, Brango had the fortune of being assigned to the reconnaissance crew for that day.

They departed Miami at noon of a bright sunny day. For three hours they flew over a calm ocean, flecked with sunlight. By then they could see the looming mass of clouds ahead, which indicated Charlie's whereabouts. Dropping from seven thousand feet cruising altitude to six hundred feet, they started getting into the eye. The sun had disappeared and the winds jumped rapidly to seventy miles an hour. For almost an hour they swung around to the west and south, feeling for the weaker side, as the winds got up to one hundred miles per hour and the rain and turbulence became terrific for about ten minutes before they broke through the inner wall and into the eye.

According to Brango, "The eye is a pleasant place! Many of them have blue sky, calm seas and air smooth enough to catch up on your reports and even drink a cup of coffee. Charlie's eye wasn't too good—big, but cloudy; still it was better than what we had just come through, so we hung around for about thirty-five minutes, watching the birds. There are usually hundreds of birds in the eye of a hurricane. Probably they get blown in there and have enough sense not to try to fly out. But not us, we want out."

Soon the decision to start out was made, and the order went over the inter-com: "Stand by to leave the eye—report when ready." This always brings the stock answer, which has become a standard joke in the squadron: "Don't worry about us mules, just load the wagon!"

The flight out was rough. Sunset was nearing, and in the storm area night falls rapidly. For almost two hours they beat their way through one hundred mile-per-hour winds toward the edge of the storm and in the general direction of Corpus Christi, their destination. The turbulence and rain on the way out were so severe that they were unable to send out messages and position reports, so someone in the crew, catching a glimpse of the waves beneath, came through with the scintillating remark that "We're still lost, but we are making excellent time."

About nine hours after they had left Miami, they landed at the Naval Air Station, Corpus Christi, Texas. An hour later they were out of their dripping flight suits and "testing the quality of Texas draught beer."

At dawn the next morning, another crew and another plane from the squadron was into the hurricane, only a few hours before it struck Tampico and then swirled inland, to dissipate itself on the mountain range to the west of that coastal city.

Shortly before the middle of September, 1948, the Weather Bureau in Washington had a long-distance call from the Baltimore *Sun*. A staff correspondent, Geoffrey W. Fielding, wanted to fly into a hurricane. The Weather Bureau arranged it through General Don Yates, in charge of the Air Weather Service, and on September 20, Fielding was authorized and invited to proceed to Bermuda at such time as necessary between that date and November 30, to go with one of the crews on a reconnaissance mission. The Air Force offered transportation to Bermuda and return at the proper time.

On the day of Fielding's call, a vicious hurricane was threatening Bermuda and the B-29's were exploring it, but it was too late to arrange a trip. On the thirteenth it passed a short distance east of the islands, with winds of 140 miles an hour. The next tropical disturbance was found in the Caribbean west of Jamaica and became a fully developed storm on September 19. As it raked its way across the western end of Cuba on the twentieth, and southern Florida on the twenty-first and twenty-second, Fielding flew to Bermuda. By the time they were ready to take off, the storm was picking up force after crossing Florida and was headed in his direction.

Not the worrying type, Fielding made notes of everything: the ditching tactics, the lifesavers and parachutes, sandwiches for lunch, the weather instruments, and the exact time of take-off, 12:03 P.M., Bermuda time. Already, high, thin cirrus clouds were seen, spreading ahead of the storm. Southward, the clouds lowered and thickened. And then the aircraft commander, Captain Frank Thompson, saw a tanker wallowing in the heavy swells a quarter of a mile below, and everybody had a look. Big seas swept over the bows of the ship and crashed on deck. The crew of the B-29 felt sorry for the men on the tanker.

"Watch that old ship roll down there," said the pilot. "Those poor guys may be in this a couple days. They make very little headway as the hurricane drives toward them. I wouldn't like to be in their place." The super fortress flew a straight course into the teeth of the hurricane and low, ragged, rain-filled clouds soon hid the tanker from view. Increasing winds buffeted the big aircraft, which now seemed like a pigmy plane in this vast wind system. They were instructed to follow the "boxing" procedure and were headed for sixty-knot winds in the northeast sector.

Over the inter-communications suddenly came the excited voice of the navigator, Lieutenant Chester Camp: "I've got them—there they are—sixty-knot winds. Bring the plane around." The plane banked in a right turn as the pilot brought the winds on the tail and shot fuel into the engines to force the plane through winds that would become more violent. So they started the first leg of the box.

The weather officer, Lieutenant Chester Evans, was seated in the bomb-aimer's position in the glass nose of the plane, practically in the teeth of the gale. In addition to keeping track of the weather, he guided the pilots by reading the altimeters to get the height of the plane above the sea. In spite of the jostling he was getting from the bouncing plane,

Fielding investigated these operations and wrote in his notebook:

"In addition to the regular altimeter, Lieutenant Evans has a radar altimeter, which works on the principle of the echo sounding machine used by ships. A radar wave is transmitted from the small instrument to the surface of the sea and bounces back again. The time elapsed between transmission and reception is computed by the gadget in feet, giving an accurate height reading. The information is passed back to the pilots who adjust their pressure altimeters. In some cases the error of the pressure altimeter measures up to three or four hundred feet in a hurricane.

"The second leg of the box started at 3:05 P.M. and was quite short, lasting only thirty minutes before the plane had run through the low pressure and then to a place where it was six millibars higher. Low gray ragged clouds increased in this sector and the ceiling lowered. On order from the commander, called Sooky by the crew, the plane went down to two hundred feet. Below, seen through a film of cloud, the water raged and boiled. Huge streaks, many of them hundreds of feet long, etched white lines on the beaten water, which was flatter than a pancake. The roaring, tearing wind scooped up tons of water at a time which, as it rose, was knocked flat again by the force of the wind. Sometimes the wind would literally dig into the water, scooping it out. From this, huge shell-shaped waves of spume would careen across the water."

At this point, someone yelled, "Sooky, take a look at the water. You'll never see this again. Wind is ninety miles an hour now." All the crew peered through the windows. The sea was absolutely flat, except for huge streaks, some of which the weather observer estimated to be at least five feet below the surface of the water. The time was 3:45 P.M., according to Fielding, who kept precise notes on everything. Instead of being thrown all over the place as he had expected, the plane was being lifted up and flopped down again in a series of sickening jolts. To stand upright called for an acrobat, not a newspaperman. He found it useless to stand, anyway. It resulted only in a hard crack on the head when the plane dropped.

At 3:55 P.M., the navigator screeched over the interphone: "It's up to one hundred miles an hour, now. Gee, is this some storm!" The rain came in torrents. "Driven by a smashing, battering wind, it hammered on the skin of the plane. The wind joined in the noise, howling and screeching outside and the roar of the engines was drowned out by the mad symphony of nature," wrote Fielding. The plane bucked and yawed but it was designed for high-altitude flying, with pressurized cabins for use when needed, and no rain came in.

They were on the third leg now and it became hotter in the plane. Everybody was sweating profusely. Fielding wrote that the "storm bucked and tossed the heavy bomber through the skies like a leaf in autumn." At 3:58 P.M., the wind was up to 120 knots. In the midst of all the noise, Fielding heard a voice on the inter-com. "How are you feeling?" came a question. "Not so good," was the miserable reply. "I wish Sooky would get the plane out of this. That blue cheese I ate in a sandwich for lunch is turning over. All I can taste is that stinking stuff." Others admitted having fluttering stomachs.

The radar operator was unable to get the eye of the hurricane on the scope. The co-pilot, Captain Hoffman, commented on the scene: "This is a big storm. It has really picked up in size." Hardly were the words out of his mouth before he yelled, "Hey, look, it's clear outside! The sun's coming through." A shaft of sunlight probed through the clouds and filled the cabin with a reassuring glow. They ran the fourth leg but there was nothing new. Fielding thought that they had seen all that this hurricane could produce in the way of violence. The radio operator got Kindley Air Base on the 42-20 frequency and learned that all other military planes in the area were warned to head for the nearest mainland base. They asked for clearance to MacDill Field and got it at 6:25 P.M. Stars appeared in a clearing sky and the plane leveled off and roared through the darkness. It was good to be able to hear the engines again. Tins of soup were opened and legs were stretched. Stomachs had settled and there was light chatter over the inter-com. The plane touched down at MacDill at 10:45 P.M. The men went to bed with aching bodies but they slept. As Fielding said at the end of his notes, "We had been eleven hours in the air, much of it in violent weather, and the constant strain tells on you."

Finally, in 1954, the so-called "hairy hop" was projected into the living rooms of people all over the country. When Hurricane Edna was headed up the coast toward New England, Edward R. Murrow and a camera crew of the Columbia Broadcasting System flew to Bermuda, and the Air Force succeeded in getting the entire group—Murrow, three assistants and one thousand five hundred pounds of camera equipment—in the front of the plane. While everybody on the crew held his breath and Murrow used up all the matches aboard and wore out the flint on a lighter, the big plane was skillfully piloted through the squall bands and pushed over into the center. The cameras ground away and Murrow asked endless questions. The eye was magnificent, called a storybook setup, clear blue skies above, the center being twenty miles in diameter, with cloud walls rising to about 30,000 feet on all sides. The return was as skillful as the entrance, through the squall bands, out from under the storm clouds and back home above blue waters and in the sunshine. The film brought to television viewers some idea of the majesty and power of a great storm.

Murrow described their passage into the eye of the storm in these words:

"The navigator (Captain Ed Vrable) asked for a turn to the left, and in a couple of minutes the B-29 began to shudder. The co-pilot said: 'I think we're in it.' The pilot said: 'We're going up,' although every control was set to take us down. Something lifted us about three hundred feet, then the pilot said: 'We're going down,' although he was doing everything humanly possible to take us up. Edna was in control of the aircraft. We were on an even keel but being staggered by short sharp blows.

"Then we hit something with a bang that was audible above the roar of the motors; a solid sheet of water. Seconds later brilliant sunshine hit us like a hammer; someone shouted: 'There she is,' and we were in the eye. Calm air, calm, flat sea below; a great amphitheater, round as a dollar, with white clouds sloping up to twenty-five thousand or thirty thousand feet. The water looked like a blue Alpine lake with snow-clad mountains coming right down to the water's edge. A great bowl of sunshine.

"The eye of a hurricane is an excellent place to reflect upon the puniness of man and his works. If an adequate definition of humility is ever written, it's likely to be done in the eye of a hurricane."

The Air Force man who made the arrangements for this broadcast, Major William C. Anderson, said that this relatively smooth flight was the best possible testimonial to the progress the hurricane hunters had made in flying these big storms, for Edna was no weakling. But he worried about it day and night until the flight was finished, for many strange things can happen. When Murrow and his crew were safely back in New York, Anderson turned in for his first good night's rest in two weeks, duly thankful that it hadn't turned out to be a "hairy hop."

14. THE UNEXPECTED

>>

"There is not sufficient room for two airplanes in the eye of the same hurricane."
—Report to Joint Chiefs of Staff

Twenty-five years before men began flying into hurricanes, it was the main purpose of the aviator to keep out of storms of all kinds. If he ventured any distance out over the ocean in a "heavier-than-air" machine, he expected to see ships guarding the route, to pick him up if he fell in the water. In 1919, when the Navy had planes ready to fly across the Atlantic, they had a "fleet" of ten destroyers and five battleships stationed along the line of flight from Trepassey Bay, Newfoundland, to Portugal via the Azores, to furnish weather reports that would help the pilot to avoid headwinds, stormy weather and rough seas, and to take part in rescue operations in case of accident.

Three airplanes, the NC-1, 3 and 4, used in this flight were designed and built through the joint efforts of the Navy and the Curtiss Aeroplane Company. These four-engined seaplanes, the largest built up to that time, exceeded the present-day Douglas DC-3 airplane in size and weight. Although sufficient fuel could be carried for a sixteen-hour flight, cruising airspeed was but eighty miles an hour. During the winter months of 1918 to 1919, plans were made by the Navy, in co-operation with the Weather Bureau, for securing as complete and widely distributed weather reports as possible for the Atlantic area immediately prior to and during the flight. Through international co-operation, observations were available from Iceland, Western Europe, Canada, and Bermuda.

From this network of reports, it was possible to draw fairly complete weather maps and to follow in detail the various weather changes which might affect the flight. There were several special features that required consideration. For example, because of the heavy gasoline loads aboard the planes, it was necessary that the wind at Trepassey Bay be within certain rather narrow limits, strong enough to enable them to get off the water, but not so vigorous as to damage the hulls or cause them to upset. Similarly, the planes would need the help of a moderate westerly wind in order to reach the Azores on the first leg of the flight, but an excessive wind would cause rough seas, making an emergency landing extremely hazardous. Thus the problem was to select a day on which reasonably favorable conditions would be encountered, and to get the planes away as early as possible, to minimize the cost of maintaining the fleet at their positions. After four days of careful analysis and waiting, the Weather Bureau representative at Trepassey issued the following weather outlook on the afternoon of May 16, 1919:

"Reports received indicate good conditions for flight over the western part of the course as far as Destroyer No. 12 (about six hundred miles out). Winds will be nearly parallel to the course and will yield actual assistance of about twenty miles per hour at flying levels. Over the course east of Destroyer No. 12 the winds, under the influence of the Azores high, recently developed, will be light, but mostly from a southwesterly direction. They will not yield any material assistance.

"Weather will be clear and fine from Trepassey to Destroyer No. 8 (about four hundred miles out); partly cloudy thence to the Azores, with the likelihood of occasional showers. Such showers, however, if they occur, will be from clouds at low altitudes, and it should be possible to fly above them.

"All in all, the conditions are as nearly favorable as they are likely to be for some time."

It is a strange fact that the Weather Bureau forecaster on this flight was Willis Gregg, who became Chief of the Weather Bureau in 1934, and the Navy forecaster for the same flight was Ensign Francis Reichelderfer, who became the Chief of the Bureau in 1938 after Gregg's death.

In accordance with this advice, the three planes departed that evening and flew the first leg of the flight almost uneventfully until the NC-1 and 3 attempted to land on the water near the Azores due to very low clouds. Upon landing, although both crews were picked up by near-by ships, heavy seas damaged the planes to the extent that they could not continue the flight. Fortunately, however, the NC-4 was able to make a safe landing in a sheltered bay, and after a week's delay, awaiting favorable weather, continued from the Azores alone, arriving at Lisbon, Portugal, on May 27.

No one at that time would have believed it possible for this situation to be reversed. Instead of waiting to be sure that the weather is favorable, planes now assigned to hurricane hunting wait to be sure the weather out there somewhere is decidedly unfavorable before they take off in that direction. But even in hurricane hunting the unexpected happens and, as in the old days, the crews are intensively trained and all precautions are taken so that they are not likely to be caught by surprise in an emergency. In a period of years there are hundreds of missions into dozens of tropical storms and, unfortunately, a few have met with disaster. So the intensive training goes on without interruption.

It seems strange but it is a fact that some men fly into hurricanes and typhoons without seeing much of what is going on outside the plane. They are too busy with their jobs to spend time looking around. In the first year some of them learn more about these big storms before and after missions than they do while flying. There are lists of reading matter to be consulted, including books and papers on tropical storms, and there are hints, suggestions, advice and warnings based on the experiences of other men. Also, they read the reports that usually are gathered from the members of other crews after their flights are finished. At the end of the season, all these pieces of information may be assembled in a squadron report, with recommendations. New men are expected to study this material. Before each flight, the crew gathers in front of a large map for a "briefing." Here an experienced weather officer shows them a weather map, points out the location and movement of the storm center at the last report, and indicates the route that seems most favorable for an approach to the storm area and for the dash into its center.

Most of this training is aimed at the development of crews that will be ready for any emergency—for the "unexpected," so far as that can be realized. Their performance in recent years shows that this special training enables them to survive most of the frightening experiences which probably would be disastrous to crews on less spectacular types of missions.

Usually there has been separate training for the men most concerned with each of several jobs—weather, hurricane reconnaissance, engineering, communications, navigation, photography, radar and maintenance. Before departure, the ground maintenance men see that the plane is in good working order and that the equipment is operating properly. At the beginning of each season, for example, some of the Navy maintenance men get the city to turn the fire hose at high pressure into the front of the plane, to see how it reacts. The effects of torrential rains in high winds of the storm are simulated in this manner. After every flight, the plane needs very thorough examination. One of the troubles is that salt air at high speed causes rapid corrosion. Salt may accumulate around the engines. Also, severe turbulence causes damage to the plane.

After the take-off, the pilot and co-pilot can see what is ahead most of the time, but for considerable intervals they are on instruments, or, as they say in the Navy, "on the gauges." They see nothing or very little of what is ahead of the plane in such cases and, the sea surface being hidden from view, they are uncertain as to their altitude until the weather officer, or aerologist, gives them a reading from the radar altimeter. Sometimes in darkness a pilot has had the bright lights turned on so that a flash of lightning will not leave him completely blinded at a time when he must know what the instruments show because of the violent turbulence that may be experienced when there is lightning. Then, too, they always have in mind that there may suddenly be torrential rain that will lower the cylinder head temperatures to a dangerous level. They must accelerate and heat the engines without traveling too fast. The landing gear is dropped to catch the wind. By using a richer mixture to feed the engines, the cooling effect may be lessened. It is always necessary to be on the alert. Altogether, it is just as important, and oftentimes more so, for the men to see the gauges than to see the weather.

Although the Air Force and Navy have different methods of flying into tropical storms, there are certain dangers that are common to both systems. Ahead of time, the pilots and others make a last-minute check to see that the crew are prepared. They also check instruments, lights, pitot and carbureter heat, safety belts, power settings, emergency equipment, current for communications and radar, and other things. In flight, the pilot does not use the throttle unnecessarily, but chiefly to maintain air speed. Actually it may be said that there are three pilots. The third one, sometimes known as "George," is the auto-pilot, which may do most of the flying, except in rough weather and in landing and take-off. Keeping the plane on course on a long flight would be very tiring otherwise. The limits of air speed vary. In the B-29's, which have been used generally for Air Force hunting, the limits are between 190 and 290 miles an hour, roughly. Air-speed readings may be affected by heavy rain. Also, increased humidity of the air will result in an increase in fuel consumption. There are numerous other items on the list of things that may cause trouble. But the pilots are highly competent and thoroughly trained and experienced before being put on the hurricane detail.

The radio operator, of course, is fully occupied and seldom has much time to see what is going on in the weather. He has two main troubles. One is static. When it is bad, all he can do is send a message blind and ask the ground station to wait. This may last for an hour or more. Various devices are used to reduce static interference but without complete success. As soon as the plane starts bouncing around, he has difficulty keying the message, not only because his body is shaking and swaying, but because it produces variations in the transmitter voltage and, on very high frequency, a drop below a certain critical voltage is likely to render the equipment inoperative.

To overcome a little of the trouble from turbulence, some radio operators in the early days tried strapping one arm to the desk, but one radio man, having just experienced a rough flight, said in his report that his arm didn't do a very good job unless he was there! Besides, he needed the arm to hold on with. More recently, it has been necessary to carry two radio men, and in fact this has become standard practice in most areas in the last year or two. It is very seldom that communications fail entirely but a plane on a storm-hunting mission that was out of contact with the ground station for much over an hour usually returned to base. Some aircraft on storm missions carry extra receivers and transmitters.

One navigator interviewed said that he is as busy as a one-armed paper hanger. He keeps track of the position of the plane by dead reckoning and by loran, which is "long range navigation," accomplished by receiving pulsed signals from pairs of radio stations on coasts or islands. It works well in the center of the storm, not so well elsewhere; in some parts of the hurricane belt, loran coverage has been poor. If it fails, the plane may go out to a point where the navigator can get a good fix by loran and do the dead reckoning from the center to this point.

Every few minutes, the navigator writes in his log a note about drift, compass heading, indicated air speed and time, and when it is rough bumps his head on the eye piece of the drift meter, the radar or something else. He takes double drift readings to get the speed of the strongest winds, figures the diameter of the eye and the exact location of the aircraft while in the eye, and passes this information to the weather officer or aerologist for his report. The duties are so numerous that the Navy usually carries two navigators "to produce pinpoint accuracy with limited celestial or electronic aids while being buffeted by one hundred-knot winds." Two are required largely because of frequent changes in heading and the nature of the winds in the Navy low-level style of reconnaissance. The Air Force uses two on daily weather reconnaissance and sometimes on storm missions.

In many respects, the weather officer, or "flight aerologist" as they call him in the Navy, is the key man on the mission. The plane is out for a series of weather reports and it is up to him to decide which is the best way to get what he wants. Within the limits of operational safety, his decisions are accepted. It is his job to keep track of the weather in every detail. He has a complicated form containing many columns in which he enters figures taken from code tables to fit the various elements—flying conditions, time, location, kinds of clouds, heights of cloud bases and tops, direction and distance of unusual phenomena, rain, turbulence, temperature, pressure, altitude, and every other conceivable detail that might be of use to the forecaster on shore. If he put this in plain language, the message would be as long as a man's arm and the radio operator might never get it off. There is an international code in figures for this purpose which makes it possible to put a very large amount of data in a brief message. And this is a continuous operation. Hardly does the aerologist get one message into the hands of the radio operator until he begins another one. It is his job to keep the pilot informed of the correct altitude. The weatherman is seated right out in front where the oncoming weather beats a terrific hubbub against the Plexiglas.

The radar operator may be one of the navigators. He keeps his eye on the scope. Many queer shapes come and go as the plane speeds along and the radar man has to know how to interpret them. He keeps the weather officer informed. Also, it may be his job to help the navigator guide the pilot around places where turbulence is likely to be excessive. Now and then, he or another crew member releases a dropsonde to get temperature, pressure, and humidity in the air between the plane and the sea.

The photographer has his troubles. Conditions are far from favorable and oftentimes impossible for taking pictures. One of his important jobs, and one that has been done exceedingly well by Navy photographers in the squadron headquartered at Jacksonville, is to get photos of the sea surface in winds of various forces from eight knots up to one hundred thirty knots. These photos are extremely useful in estimating the force of the wind by watching the effects on the sea.

In addition, there is an engineer. He looks after the over-all operation of the plane and watches the many instruments on the panel. Usually he is a man of long experience who has worked up from crew chief. He adjusts power to fit the fuel load. If an engine catches on fire, he knows how to put it out. If a bail-out is imminent, he is the man on the job Sitting behind the second pilot, he has his restless eyes concentrated on the mechanical equipment. All of these men on the plane work as a team, any of them being ready to help somebody else in an emergency, and alert and resourceful to to take quick action when the unexpected happens, and it often does.

The crews are usually organized as follows: The senior pilot is in command—in the Navy he has the title of "Plane Commander" and the other pilot is the "Co-Pilot." In the Air Force the man in charge is the "Aircraft Commander" and his assistant is "Pilot." In any case, both of these men are heavily engaged in keeping the aircraft under control when the weather is rough. The pilots, together with two other men, the engineer and the crew chief, keep the plane in the air, though these latter two jobs may be combined, in which case the crew chief has an assistant—a flight mechanic.

Under the crew chief, or crew captain, there is one exceedingly important duty—watching the engines. On each side a man looks constantly for signs of trouble—oil leaks, fire, or whatever. These two men are sometimes called "scanners." White smoke or black smoke, as the case may be, on issuing from an engine signals a dire emergency. It may be only one or two minutes from incipient fire to explosion, and action must be immediate to put the fire out or correct other troubles. It is a very definite strain on the scanners to be alert every instant on a long flight, and various members of the crew may be rotated on these jobs. On routine daily reconnaissance in non-hurricane weather, the Air Force flights are long and some of the men feel decided relief on taking a hurricane mission, which is rougher but usually much shorter.

With this training and organization of the crews, most of the emergencies are met quickly and efficiently. Now and then, the unexpected happens, however, as is evident in the following instances.

In September, 1947, a number of missions by the Navy and Air Force had secured data in Hurricane George and the big storm was headed ominously toward Florida. An Air Force crew was in it on September 16 and had been in trouble. There were gasoline leaks, several fires, and engines acting up. They decided it was an emergency and set course for MacDill Field. Everything went well until they approached the field for a landing. There, in the middle of the runway, sat a big turkey buzzard. In the twinkling of an eye, when they were only fifty feet away, the great bird took off and smashed into the leading edge of the right wing. The impact made a sizable dent and the wing dipped. After six tries, the pilot skillfully got the plane down without an accident but the crew was more upset by this bird than by the average hurricane.

Sometimes the unexpected leads to disaster. One of the most unfortunate of these incidents occurred at Bermuda in

1949. There was a report of a disturbance in the western Caribbean on November 3. It was late in the season, but a few very bad hurricanes have struck in this region in November, so the forecasters at Miami asked for reconnaissance and the request was passed to the Air Force at Kindley Field, in Bermuda. It was afternoon when the message came. A B-29 with a crew of thirteen men was cleared for a flight through the storm area and thence to Ramey Air Force Base, in Puerto Rico, where they were to spend the night.

The plane took off at 6:17 P.M., Bermuda time, climbed to ten thousand feet and leveled off. Almost immediately the crew saw an oil leak in the No. 1 engine and it was feathered. The radio operator got in touch with the tower and airways and the aircraft commander prepared to return to the field. The pilot brought the plane over the island and reported at four thousand feet, descending. But just at that time a Pan American Stratocruiser was cleared to land. The B-29 circled and reported at one thousand five hundred feet at a distance of seven miles west of the island. Next the plane was four miles out, coming straight in at one thousand feet and was cleared to land on Runway 12.

There was a gusty cross wind and there were scattered clouds at one thousand feet. The plane then reported that it would pass over at one thousand feet and get lined up, but almost immediately said to disregard the last message. One-half mile away, the flaps were raised, the landing gear was let down, and power was applied on the three remaining engines. The plane made a left turn which became steeper and altitude was lost rapidly until the left wing hit the water. This was a quarter of a mile offshore. Fire broke out as the plane hit the water and rescue boats rushed to the scene. Only three men escaped, two of them miraculously through a hole in the fuselage, as was determined by a Bermuda diver who went down sixty feet in the water to examine the wreckage. The other man, captain of the aircraft, was pulled out but died later in the hospital. It was the two radar men who were fortunately in a position to get out through the hole in the fuselage and both survived.

In this incident at Bermuda the plane was not being affected by a storm. It is an amazing fact, in consideration of the very large number of weather missions flown by the Air Force after World War II, that their first plane to be lost while on reconnaissance in a tropical storm was in 1952. On November 1, a B-29 left Guam to fly into a typhoon called Wilma. The crew of the superfort was instructed to penetrate the storm, report by radio, land at Clark Field in the Philippines, and be prepared to fly through the typhoon again on the following morning. The same day, however, radio contact was lost. Seventeen rescue planes and numerous surface vessels searched the typhoon-torn waters near Samar Island for survivors without success. Natives on the island of Leyte reported that a four-engined plane was seen flying low in that vicinity but the report could not be verified.

The squadron to which this plane was assigned had made more than five hundred reconnaissance flights into typhoons between June 1, 1947, and the date on which it was lost.

Lieutenant A. N. Fowler, an experienced Navy pilot, was the man who said that a hurricane flight was like going over Niagara Falls in a telephone booth. Describing one of his most dangerous trips, he told a newspaperman:

"I have seen the hurricane-swept sea on many occasions, but it never fails to impress me in exactly the same way. It would be sheer turmoil, like a furious blizzard. While experiencing the jarring turbulence, the heat and drumming of torrential rain which seeps in by the gallon and tastes salty, the inside of a hurricane can be like a bad dream. Like having been swallowed by an epileptic whale, or going over Niagara Falls in a telephone booth."

On a less serious note but illustrative of the unexpected, there is the tale of the Navy crew and the hot water. They took off in a Privateer to fly into the center of a hurricane, each member of the crew having been assigned certain specific duties, as is always the case on these missions. The radar operator, among other jobs, was given the coffee detail. After a considerable period of moderate to heavy turbulence, with heavy rain leaking into the plane until everybody was thoroughly soaked, they broke into the clear in the eye of the hurricane, about twenty-five miles in diameter. The weather officer was busy with the coding of his latest observation, the radio operator was sending two messages that had accumulated, and the navigator was figuring the position of the eye and computing a double drift for wind. The co-pilot had the controls and was flying around the eye, preparatory to a descent as soon as the coffee had gone around.

The pilot called for coffee. The radar man dragged out two jugs, both still hot, and began to pour. He threw the first cupful under his seat and poured one from the other jug. Then he saw that he had brought two jugs of hot water and no coffee. "What the heck!" exclaimed the weather officer. "Why, you poor ----!" The navigator's words were scathing. He said that, according to the Bible, Noah was tossed overboard for less reason.

From the very beginning of reconnaissance, these missions have been co-ordinated according to instructions issued by a trio who serve on the Joint Chiefs of Staff and also on the Air Co-ordinating Committee. Today the men are Brigadier General Thomas Moorman of the Air Force, Captain J. C. S. McKillip of the Navy, and Dr. Francis W. Reichelderfer, Chief of the Weather Bureau. There have been no serious accidents on the Atlantic side when planes actually were in hurricanes and there was no confusion in assigning planes until September, 1947. The men on the Committee at that time were Brigadier General Donald Yates and Captain H. T. Orville, in addition to Dr. Reichelderfer. They co-ordinated many operations in addition to hurricane reconnaissance and all had had long experience in aviation. Dr. Reichelderfer was formerly in charge of weather operations in the Navy, after long experience at sea and in the air. He was weather officer for Hindenberg on his flight around the world in a dirigible.

On September 18, 1947, the committee was surprised and alarmed by a report of reconnaissance. An Air Force plane out of Bermuda flew into a big hurricane which was moving west-northwest to the south of Bermuda and, after a rough time in the outer parts of the storm, finally found its way into the eye. Immediately they saw a Navy Privateer flying around in the center, also on reconnaissance, and they got right out of the eye and returned to base. There they made an official protest that there is not sufficient room for two planes in the center of the same hurricane. New instructions for co-ordination were issued immediately to all concerned. It is not surprising that this has happened on at least two other occasions, once with two Air Force planes and on another occasion with a commercial airliner.

In 1953 there was another bad accident, but not directly in a hurricane area. It resulted from a moderate hurricane named Dolly, which came from the vicinity of Puerto Rico on September 8 and moved toward Bermuda with increasing intensity. On the tenth, aircraft in the center estimated the highest winds at more than one hundred miles an hour, but on the eleventh it weakened and passed directly over Bermuda. There were strong gales at Bermuda, although the storm was diminishing in force so fast that no serious damage resulted.

On the tenth an Air Force plane from Bermuda flew into the hurricane. A Weather Bureau research man, Robert Simpson, went along to follow up on some studies he was making of the circulation at high levels in tropical storms. He reported:

"Dolly was an immature storm with most of the cloudiness concentrated in the northern sector. On the south and west sides, clouds rose only to around seven or eight thousand feet near the eye, except along the spiral rain bands which encircled the eye. The plane first investigated conditions at one thousand five hundred feet in the eye, where it was observed that there was a huge mound of cloud near the center with a moat or cloudless area which encircled this central cloud and separated it from the walls of the eye."

After this low-level exploration, the plane climbed to 29,500 feet, completing a spiral sounding in the eye. At this elevation or slightly lower, a complete navigation of the storm area was made, with dropsondes being released in strategic quarters, pressure and temperature gradients being measured along the track of the plane. There were two outstanding things observed during this flight at high levels: first, the sheer beauty of the storm itself, which could be viewed in excellent perspective, insofar as the cloud forms were geared to the wind circulations over hundreds of miles surrounding the eye. The only obstructions to vision at this elevation were the tall cloud walls which rose from the northern side of the eye. The second was a strong cyclonic circulation near thirty thousand feet over the eye itself which was surprising. Most theorists had figured that the cyclonic circulation would cease at high altitudes and possibly at very high levels become anticyclonic.

Simpson continued:

"By the time the plane had returned to Bermuda it was evident that Dolly was bearing down upon the island itself and that everything had to be evacuated. All of the planes were flown out to the mainland and the buildings battened down for the big blow. I spent most of the time in the weather station with my eyes glued to the radar scope. As the storm approached, and the winds rose, one rain band after another passed over the station, each with evidence of a little more curvature than the preceding band.

"Finally, the scope indicated a circle with a five-mile area free of any radar echoes. It was bearing down directly upon Kindley Field. Oddly enough the pressure had not begun to fall and the wind was holding steady. Another odd thing was that during the reconnaissance the eye had been twenty-five miles in diameter. However, this eye was only four to five miles in diameter. The eye arrived, the rain stopped and then resumed as the eye passed over the station, yet the pressure only leveled off briefly and the wind only subsided slightly without shifting. We had been tricked! This was not the real McCoy, it was a false eye. Subsequently, two other false eyes appeared on the radar scope and we had about decided that the storm had no organized central circulation left when the real thing finally showed up on the scope, still twenty-five miles in diameter."

In the reconnaissance of Hurricane Dolly, many feet of radar pictures were made of the spiral bands of the storm. When it became clear that all planes would have to be flown to the mainland because of the approach of Dolly to Bermuda, the film pack used on the reconnaissance was left in the plane so that additional pictures could be made on the flight back to the mainland. Not only was this done, but also an additional eye dropsonde was obtained during the trip to the mainland. It was agreed that as soon as the plane returned to Bermuda after the storm had passed, the film and additional records would be mailed to Washington.

On its flight from the mainland while returning to Bermuda, the plane exploded in mid-air 150 miles off the coast, near Savannah, Georgia. It had the records, the radar film, the dropsondes taken in the eye, and other data. In this case, the No. 4 engine had "run away," throwing its prop, which struck Engine No. 3, and the latter exploded. The plane fell out of control. Eight of the crew were rescued but none of the records or data of the reconnaissance was saved. This plane, however, was not on a storm mission at the time.

The unexpected appearance of a small eye on the radar scope is not uncommon. The Navy's instruction to its crews says: "During the final minutes of the run-in, radar may prove to be more of a hindrance than a help. There can be a number of open spots close to the true eye which might appear as eyes on the radar screen. You should not chase these false eyes!"

Out in the Pacific, the typhoon chasers say: "False eyes are often found in weak storms and care must be taken not to confuse them with the true eye of the typhoon. On the radar scope they may present an appearance much like the true eye but will not remain on the scope for any length of time. By continually scanning the suspected eye with several sweeps, the radar observer will see that the false eyes are surrounded by fuzzy cloud formations rather than a heavy ring of cloud characteristic of the eye."

When Hurricane Carol of 1954 was approaching the New England Coast, the last penetration was made by a Navy plane with Lieutenant Commander R. W. Westover as pilot and Lieutenant C. W. Hines as co-pilot. On the way into the storm circulation, Hines was telling Westover about his family's experience in the New England hurricane of 1938.

The family residence was on Cape Cod. It was blown into the water and drifted until it lodged against a bridge, obstructing navigation. Finally, it was necessary to dynamite the wrecked house to clear the channel. The Hines family rebuilt their home and took out hurricane insurance. They carried the insurance until June 1, 1954, and then let it lapse.

As the recco plane flew into the center of Carol on August 30, the crew was watching a Moore-McCormack ship in the stormy seas below and sympathizing with the people on board who were suffering such rotten weather, but Hines was saving his sympathy for his family on Cape Cod. He was sure that Carol was going to blow their home into the water again, and afterward he learned that it did.

Although Carol of 1954 received a great deal of publicity because of death and destruction in New England, Westover, who also flew into Hurricane Carol of 1953, says that it was a much more violent hurricane than the one in 1954. The first Carol was so bad that only one low-level penetration was attempted. His crew recorded pressure 929 millibars in the center—about 26.80 inches—and they recorded 87½° drift. But fortunately the earlier Carol remained out at sea throughout its course.

Hurricane Hazel, later in 1954, gave another Navy pilot, Lieutenant Maxey P. Watson, an experience of the same kind that Lieutenant Hines had. The storm was approaching the coast of South Carolina when Watson flew his plane into it and he saw the center passing inland not far from the town of Conway, which was his home.

Hazel was responsible for other unexpected incidents here and there during its ravages from the Caribbean to the northeastern part of the United States. One case was on a Navy plane commanded by Lieutenant G. J. Rehe. Watson was the pilot on this trip, also. They took off from Puerto

Rico and flew into the storm as it was turning northward and passing out of the Caribbean.

Up to that time, Hazel was not much of a storm. Westover flew into it after it passed Grenada and found that it was not a well-organized cyclone. Rehe had gone into it on the first penetration and reported winds of eighty-five knots. Westover found the area almost cloudless but ninety-knot winds in one area. However, after its northward motion began, it was a very dangerous wind system, which was responsible for the only injury to a Navy crewman in their many flights into this particular hurricane.

Because of the severe turbulence that had developed quickly in Hazel, all the crew members on this flight were fastened in with safety belts, as is usual in such cases, but the photographer wanted to get up and take a picture. So he got out of his safety belt and had another crew member unfasten himself and hold him while he took the picture. In the sudden very violent turbulence, both were thrown against the overhead. On his descent, the photographer caught his arm between the cables and the fuselage and broke his shoulder blade. The other crewman was knocked unconscious.

Out in the Pacific, an Air Force pilot, Captain Leo S. Bielinski, had an experience which induced him to go to great lengths of experiment and ingenuity in an effort to find an easier way to track typhoons and hurricanes. It was in May, 1950, when a typhoon called Doris was growing to maturity while near the island of Truk and showed signs of changing its path, threatening the base at Guam. On May 8, an RB-29, under the command of Captain Cunningham, was sent out to penetrate the storm. Bielinski went along.

At that time Leo had a fine wrist watch in which he took much pride. A man in uniform has few things that are different from the other men, but Leo secured an expression of individuality through a wrist watch. He bought a very special one for a hundred dollars and admits that he frequently looked at it when he really didn't care what time it was.

On this first trip into Doris, everything went smoothly. The crew members were instructed to land at Iwo Jima, when another plane would take over. But before landing they found that the hydraulic system needed repairs. Cunningham brought the plane down skillfully and they worked all night making repairs with parts salvaged from another plane on the field. The plans were changed and they were assigned to the next mission. The next morning they were airborne again for another penetration. This confirmed the northwest movement of Doris, which would take the most violent winds away from Guam, so they returned to Iwo Jima, well worn-out by two successive flights and thinking about a little rest, when Commander Cunningham received the following message: "Unable to get relief; request you make afternoon fix." So the same crew turned around and started the third mission. The other two flights into this storm had been uneventful, they were tired, and Leo didn't bother to fasten his safety belt.

Wham! Suddenly he found himself floating in the air around the cockpit. Before he could get his bearings, he was thrown violently against a bulkhead and slowly came to the realization that the bits of junk dangling in his face were the remains of his hundred-dollar wrist watch. This bothered Bielinski more than a broken arm or a twisted vertebrae. He started studying typhoons with a determination to find a better way to keep track of them. The results are described in Chapter 17.

In other ways the unexpected can be serious. One experience is cited by Captain Ed Vrable, who was navigator on a flight into a hurricane in 1953. After a careful approach, the aircraft suddenly popped into the eye, but it was only about eight miles in diameter. It was not easy to circle a superfortress in this small eye. At one point, the turning arc was a little too broad and the aircraft edged out into the winds on the border. It was instantly tossed back into the eye, almost upside down, and he had the worst fright of his career in the reconnaissance business. But the pilots made a skillful descent until they managed to get the plane into the correct attitude and finished the flight.

In Hurricane Edna, in 1954, a crew of hunters in a WB-29, in command of Captain Charles C. Whitney, had an unexpected duty. They had spent part of the morning and the afternoon of September 14 in the eye of the hurricane. They flew in tight little circles, dodging the wing-shuddering winds on the periphery. Because the Weather Bureau forecasters were afraid of a repetition of a sudden speed-up like that of Hurricane Carol two weeks before, they had asked for a continuous watch. Captain Whitney and his crew were in there for nine hours.

And then, with gas getting low, they ran into the unexpected. Some eleven hours after take-off from Bermuda, the aircraft picked up a radio message that the Nantucket lightship, torn from her moorings by terrific winds, was adrift and at Edna's mercy. The WB-29 plunged into 145-mile-an-hour winds in search of the vessel.

Picking up the lightship by radar, the weather plane shepherded the hopelessly lost ship, remaining overhead until a Coast Guard rescue plane arrived.

Waves seventy feet high seemed to toss the stricken vessel into the air to meet the low-flying aircraft pressed down by Edna's raging winds. It felt, the crew said later, as if the plane were dancing on her tail.

With the arrival of the relief plane, the WB-29 turned landward. After sixteen hours in the air, and with the gas gauge hitting the low side of the dial, the weather plane made a landing at Dover, Delaware.

According to the Air Force, "This flight was one of the most dramatic missions in peacetime Air Force history."

15. FIGHTING HAIL AND HURRICANES

>>

"I wield the flail of the lashing hail,
And whiten the green plains under;
And then again I dissolve it in rain
And laugh as I pass in thunder."
—Hebert

At first thought, most people would say that fighting hail has nothing to do with hunting hurricanes, but in one instance it did. It is an interesting story which shows how men will take risks in trying to control the weather. The story ends with one man giving up his life in a sensational adventure with a mysterious conclusion.

Destructive storms are not very frequent in any one place but most people are under the impression that they are. They are apt to remember bad weather and forget about the good. Losses of life and property and failures of plans and business enterprises are caused by storms or the wrong kind of weather and such things are impressed on their memories. When rain is needed, it may fail altogether or come in such quantities that fields and roads are washed out and there are floods in the rivers. A thunderstorm brings rain but sometimes hail comes with it, destroying crops and damaging property.

People have tried to overcome these bad effects of the weather in many ways. Irrigation has long been practiced in regions with scanty rainfall. Air conditioning affords relief from excessive heat. In many other ways, some foolish and some dangerous, men have tried to influence the weather. An interesting case of this kind which appealed to the imagination of people in many countries started near the beginning of the present century. It was an international battle against hail. Its origin was in the vineyards of Italy. Hail had done great damage there year after year, and finally an Italian got the idea that he might destroy hailstorms by shooting into them when they were just beginning.

In those years, cannon were used in battle. Loaded with big charges of gunpowder, these cannon hurled solid, heavy balls at enemy cities, forts, fleets, and troops. In time of peace, there were many of these old cannon around, serving no useful purpose, and the Italian had no trouble in getting one to try on hailstorms. But he was not permitted to use a cannon ball. It might have crashed into a neighbor's house or killed somebody in the vineyards. So he loaded it with gunpowder and fired it at the storm cloud, hoping it would create a disturbance in the atmosphere and weaken the hailstorm.

It is an amazing fact that the vineyard of this Italian was damaged far less by hail than those of any of his neighbors, and the next year others tried firing a cannon with similar success. They became expert at it and learned how to load a cannon so that it cast a big, whirling smoke ring into the thunderstorm cloud. The news spread to other countries and in two or three years there was a lot of hail shooting in different parts of the world. So they held an international hail-shooting congress where they exchanged ideas and narrated their experiences. By the time the second world congress on hail was held, a great deal of uncertainty had developed. It seemed that the first hail shooters had begun work at a time when it just happened that there was much less than the usual amount of hail. Also, there were explosions and people were hurt. One man was killed and another had an arm blown off. After a few years, all the hail shooting ceased.

Even today, there is a good deal of mystery about the formation of hail and many people think there are ways of preventing it or causing the storm to make little hailstones instead of big ones and thus having much less destruction. Hail causes many millions of dollars worth of damage every year in the United States and almost any effort to reduce the losses seems to be justified.

Scientists believe that hailstones are very small in the beginning but grow in size as they go up and down several times in the thunderstorm clouds. Even in hot weather, it is very cold in the top layers of one of these great clouds. Raindrops freeze and in falling gather more water or snow in these high regions. Soon they are caught in rising air currents and carried up into freezing temperatures again. On each trip up and down, another layer of water or snow gathers on the outside and is frozen. At last the multi-layered stones become so heavy that they fall to the ground, in spite of rising currents, and as they leave the cloud they come down with great rapidity and may beat crops to the ground, batter automobiles, break glass and bruise and sometimes kill livestock. A hailstone the size of a baseball falling many thousands of feet is a very dangerous thing.

For many years after the hail-shooting experiments, it was thought that nothing could be done about it except to carry hail insurance. Then, shortly after World War II, scientists of the General Electric Company announced that they had conducted some successful experiments in controlling the weather and this led to efforts to control rainfall, prevent hail, and stop hurricanes.

The man who started this new effort at weather control was Vincent Schaefer. He observed the weather on top of Mt. Washington, in New Hampshire, a place where it is very cold and windy in winter. The observatory is fastened to the solid rock of the mountain top by steel cables, to keep it from being blown off. Vast quantities of ice accumulate on the building. Snow comes down in great quantities at times but is generally carried by high winds which have reached terrific speed, on one occasion going up to 231 miles an hour. Conditions there are in some respects like the weather in the top of a big thunderstorm.

One of the peculiar things that happens up there on Mt. Washington and in the top of a thunderstorm is the formation of liquid water droplets, which are colder than freezing but they do not turn to ice. These droplets are said to be supercooled. Schaefer found in his experiments at General Electric that a small pellet of dry ice, the size of a pea, when dropped into air containing a cloud of supercooled water droplets could produce untold billions of small ice nuclei. So he carried some dry ice up in an airplane and dropped it into the top of a cloud with supercooled water droplets, and a trail of snow was seen falling from the bottom of the cloud. Many others tried the same experiment and some had similar results. The snow turned to rain as it came down to warmer levels, and the process was called "rainmaking."

There is one disturbing fact. Before dry ice will work on a cloud, it must be very near the point of making rain without any outside help. But many of the rainmakers believe that dry ice makes more rain fall or causes it to fall sooner than it would otherwise. Thus, as the cloud moves along, the rainmaker may be able to cause a shower in a certain place, whereas the cloud might have moved far away before it began to rain. In this story the important point is that some of the experimenters believe that dry ice or some other chemical will cause the rain to fall but will make it much less likely that nature's process will develop to the point of producing hail.

The news of all this rainmaking in the West aroused intense interest on the part of a young man named Gordon Clouser. He thought he might be able to prevent hail, and if he succeeded, he might stop tornadoes. In the Midwest there is an old story about a farmer who knocked the life out of a tornado by hitting it with a two-by-four. On hearing this story, many people have gotten the idea that the government might destroy a tornado by gunfire. More recently there have been serious proposals that these vicious local storms with funnel clouds and violent winds be destroyed by guided missiles. There is no evidence that any of the plans offered so far would be successful in breaking up hailstorms or tornadoes, but they are extremely small when compared with hurricanes, and the government has received thousands of proposals that these great storms be wiped out or rendered harmless by gunfire.

Behind most of the suggestions for killing hurricanes is the idea that they begin as small whirls in the atmosphere and go through early stages of growth to the size of a tornado or a thunderstorm, and if they could be hit with great force in a vital place while small, they might die out. On this assumption, there have been a great many proposals that the Navy send battleships into the hurricane area to search for incipient hurricanes and fire broadsides into them. No test of this kind has been made for two reasons. The hurricane region is so large that the entire Navy would be insufficient for such a patrol. On the other hand, there is not a shred of evidence that hurricanes begin as small storms like tornadoes or thunderstorms. Actually, they seem to develop as mildly disturbed weather over an area of thousands of square miles. The experts say that shooting at the weather in such a large region would certainly be futile. After the World War II, the atom bomb stimulated some new ideas and thousands of letters were written to the government about knocking a hurricane out with an atom bomb at the right time and place.

When the New Mexico atom bomb was exploded, the weather was bad, with rain in torrents, strong winds, lightning and thunder. Afterward, the weather was much better and this led to a lot of speculation. The fact is, however, that the scientists waited until the weather improved before they exploded the bomb; hence neither the bad weather nor the improvement could be attributed to the explosion.

Before the tests at Bikini in 1946 and Eniwetok in 1948, the scientists received numerous letters, warning them that the explosions would start storms and might cause a typhoon. But the effects of explosions of this kind are soon over, while the forces that maintain a hurricane or typhoon must be applied continuously day and night for a week or two, to keep one of these big tropical storms going in full fury. One of the scientists who witnessed these tests estimated that it would take a thousand atomic bombs at any moment to equal the energy of motion in a hurricane. No scientist has figured what would happen if one thousand atomic bombs were exploded at one time in a storm area!

After a year or two of rainmaking with dry ice and another chemical, silver iodide, the conviction grew that it would be possible to kill a hurricane by dropping some of this material in a vital spot. Some of the bolder students of weather control actually tried it. One of them was Gordon Clouser.

Just what he did when he flew into the storm and what happened to it afterward make a mystery, for he gave his life in the effort. It is a good example of the fearless activities of the hurricane hunters.

Gordon Clouser was born in 1912, in Gibraltar, Pennsylvania. He grew into his teens as an active, good-looking boy with many diverse interests. Quick to learn, he finished high school at fourteen. His family moved to New Mexico, where he worked several years as a surveyor, then took two degrees at the University of New Mexico. After that, he had many activities—teacher, librarian, writer and director of plays. He made a movie, composed music, wrote poetry, was in the Air Corps reserve one year, taught meteorology and aeronautics at Boeing Aircraft in Seattle for a year and a half. He learned to fly in Idaho and then was a teacher in Junior College in Yakima, Washington.

It was 1950 when Gordon became excited about the work that was being done in rainmaking in many parts of the country. By April of the next year, he had moved to Plainview, Texas, and had begun to organize airplane operations to prevent hail on the high plains of the State. Having developed his own secret formula for the chemicals to be dropped into thunderstorm clouds, he experimented in his car, in airplanes and in the home freezer. Once he came home for dinner, carrying some denim to be used in connection with an experiment, and his wife discovered that he had taken all the food out of the freezer so he could drop chemicals in it, to see what might happen in the atmosphere. When he asked what they were having for dinner, she replied, "I guess it will be frozen denim."

The year 1951 was not an easy one for Clouser. The thought of preventing hail was new to most people and he had some difficulty in getting enough money to finance the necessary plane operations. He asked farmers for twenty to forty cents an acre for protection from hail and compared this cost with the much higher rates for hail insurance. But, he argued, the prevention of hail would lower the insurance rates, which are based on the frequency of such storms in any area and the amount of damage done.

To prevent hail, Gordon and his pilots flew into and over thunderstorms, to see if they contained hail in dangerous sizes and, if so, they dropped his secret chemicals into the tops of the clouds. This is called "seeding" by the rainmakers. Gordon was sure that he was preventing hail damage from the clouds they seeded. By 1952 he had nine planes at his command. In that year, from June 1 to October 1, they checked 421 thunderstorms and found ice in dangerous sizes in eighty-two of them, which were seeded. He reported to the farmers that there was no appreciable hail damage from any of them and there were no complaints on that score.

During this time he was watching the reports of tornadoes and getting the Weather Bureau's forecasts and warnings. On May 26, he heard a prediction of tornadoes in an area which included the two counties where he was working to prevent hail. Without regard for the danger of flying among thunderheads in tornado weather, his planes were in the air for a total of nearly ten hours that day, seeding clouds that looked dangerous. That night, a half hour after the last of Gordon's planes landed, the Weather Bureau issued an "all clear." There had been no tornadoes in either county. Gordon said, "We can't prove that we prevented a tornado—maybe none would have formed anyway—but we do know that conditions were right for one, and we changed those conditions."

For a man of Clouser's adventurous spirit, this was just a side issue. He occupied much of his spare time studying hurricanes and making plans for the day when he would be operating a large company to kill these storms before they reached the Coasts of the United States. He hoped to have his main office in San Juan, Puerto Rico, with planes stationed also at Pensacola, Florida, on the coast of Mexico, in Cuba, and at two or three other strategic places. He would get the government reports, talk to the weather men, and at the right time drop a mixture containing his secret formula into the eye of the storm or some other vital spot that he would find by flying above the storm clouds and studying the wind circulation.

His wife, Olive, took this philosophically. With their three children, she was living at Norman, near Oklahoma City. Like the wives of most adventurous pilots, she knew that any one of these trips might be her husband's last. She encouraged him in his hail prevention but worried about tornadoes, and especially hurricanes. She knew that they form and move over vast sea surfaces on which the winds impress violent motions, a deadly place for a man to land when in trouble. After Gordon flew into the tornado clouds in May, he came to Oklahoma City by bus and called her on the phone to come and get him in the car. Instead of going home, he asked her to drive him to the Weather Bureau Office at the airport, where he checked on the reports to see if they knew what had happened to the tornadoes. Then she found out what he had been doing and heard him talking about hurricanes.

Olive had something special on her mind. She wanted to paint the kitchen yellow, but he was against it. She tried to get a compromise. If he was going to fly into tornadoes and other storms against her advice, why not paint the kitchen yellow, even if he didn't like it very much? He offered strong objections and she put it off for a while.

In the meantime, Gordon was in trouble. September of that year—1952—was very dry in Texas. The farmers in Floyd and Hale Counties in that state got the idea that his agitations against hail had prevented rain. Anyway, he was out of work, for, as he said, "There is no point in a hail-busting business when there are no clouds." A delegation of farmers called on him to protest his activities. They said that he and his men had deprived them of rain and they were going to lose a lot of money.

Gordon convinced them that his work on the clouds earlier in the year had nothing to do with the drought. He pointed out that only 82 out of 421 storms had been seeded; therefore, 339 of them had acted exactly as nature had intended. Besides that, he showed them news reports that nearly all of Texas was dry, some parts being much drier than the counties he was working. They went home satisfied, but Gordon had time on his hands, with no thunderheads or clouds to work on. So he gathered data on hurricanes and spent a good deal of time at home, making experiments in the freezer. He wanted to work on big storms. The little ones in Floyd and Hale Counties gave him trouble. All rainmakers know that it is possible to seed a cloud and have rain on the farm or ranch of a man who refuses to pay for seeding, and have no rain on a farm next to it, owned by a man who has paid for the service.

October came and it proved to be the driest month for the country as a whole since weather records began. All the rainmakers were in trouble and the "hail-busters" were out of work. Gordon sat at home, listening to the radio and working on his formula. He and Olive talked about many things but neither mentioned hurricanes or yellow kitchens. Then on Tuesday, October 21, Gordon left for Plainview. The next day he heard a news report from Lubbock that there was a hurricane in Cuba, moving toward the United States. On Wednesday he left for Florida in a Luscombe plane, saying nothing to anybody except Bill and Pauline Seirp. Bill was not a pilot but Gordon had been teaching him to fly.

Knowing nothing about the trip to Miami, Olive was having the kitchen painted yellow and wondering what Gordon would say when he came home from Plainview. That was on Thursday. On Sunday, the twenty-sixth, she and the children had a late breakfast but managed to get to Sunday School and remained for church service. During the hymn at the beginning of the service, there was a long-distance call for Olive from Plainview. Gordon was lost at sea. Later in the day, she heard the story in full.

Gordon was not satisfied with the plane. When he reached Florida he tried to get one better suited for storm work. He had plans for building a special plane for the purpose but now he was anxious to get into the hurricane. It might be the last one of the season, he thought. It had done a great deal of damage in Cuba. He went to the Weather Bureau Office in Miami and got the latest information on the position, strength and movement of the storm. At 3:45 P.M. (October 25) the center of the hurricane was about seventy-five or eighty miles east of Miami when Gordon took off in his Luscombe plane. At 8:56 P.M., a radio station in Miami picked up a message from him, saying that he was fifty or sixty miles east-southeast of Miami, still in the edge of the storm. The radio station talked with him for twenty-six minutes as he flew toward Miami, making poor headway against the winds. The last message was, "Out of fuel—descending—give my love to my wife and family."

The Civil Air Patrol and the ships and planes of the Coast Guard searched the area for forty-eight hours without finding any trace of the missing man. Olive went to Miami and did her best to keep the planes looking for him. Whether or not he had any effect on the storm will never be known for sure. The weather forecasters in Miami did not think so. But the hurricane soon afterward took an erratic course. It was destructive early on the twenty-sixth as it turned into the Bahamas, then lost force, and turned northward. The official report of the Weather Bureau said that "it moved northeastward thereafter as a disturbance of no great violence."

The uncertainties and the tragedy in this case brought to mind the Savannah storm of 1947, which Gordon may have studied. It began far to the southward, near the Isthmus of Panama, early on the ninth of October. On the eleventh, it crossed the extreme western end of Cuba, and on the twelfth passed over southern Florida. From this time on, its course was very unusual. Reconnaissance planes followed it going northeastward over the Atlantic until the night of the thirteenth, when it was east of Wilmington, North Carolina. Early on the fourteenth, a plane got into the storm area and found it moving southwestward. With considerable force it struck Savannah, Georgia, early on the fifteenth, causing about two million dollars' worth of damage. Citizens of Savannah and some of the city officials complained to the government for causing the hurricane to strike the city.

At about the time, or just before the hurricane changed its course abruptly to the southwest, military planes had carried out an experiment in dropping dry ice into its upper levels. There was a great deal of discussion in the press. At first it was said that the dry ice had caused the storm to take a new course, but after the Savannah complaints were heard, little more was said by the military about the experiment and it remains something of a mystery. Few scientists believe that dry ice could have such an effect on so large a storm. Actually, there were few observations in the storm area during the night of the thirteenth to fourteenth and precise information about the time and nature of the change of course was not available for an investigation. It belongs in the same class as the Clouser storm.

16. CAROL, EDNA, HAZEL OR SAXBY!

>>

"But I know ladies by the score
Whose hair, like seaweed, scents the storm;
Long, long before it starts to pour
Their locks assume a baneful form.
—Hebert

At the end of August, 1954, when the hurricane named "Carol" devastated Long Island and the southern coast of New England, it did a tremendous amount of property damage, principally on the shores of Rhode Island and southeastern Massachusetts. There was sharp criticism of the weathermen and the hurricane hunters. People claimed that the warning came only a few hours ahead of the big winds and the high storm tides. The weathermen answered that there really was no delay on their part in giving out the warning. They said that the hurricane hunters had been tracking Carol for several days and everybody had been warned that it was on the way. The hurricane simply started to move with great rapidity during that final night and there was no way of getting the warning to large numbers of people that early in the morning. It was after daylight when they got out of bed and turned on radio and television.

Of all the criticism, the sharpest and most prolonged was about the name of the hurricane. A newspaper in Massachusetts—the New Bedford *Times*—ran an editorial saying that it was not appropriate to give a nice name like Carol to a death-dealing and destructive monster of this kind. Other newspapers and many citizens here and there around the country joined in, partly in complaint and partly out of curiosity and the wish to get into the argument. A New Orleans woman wrote to the editor of the New Bedford *Times* that she would rather a storm would hit her house nameless than to run a chance of having it named after one of her husband's old girl friends. Other women were incensed because storms had been called by their given names. The weathermen had a good explanation, but not many people seemed to sympathize with them. Persons who suffered losses of property were the most critical, saying that the name Carol gave the impression that the storm was not dangerous and that its winds and tides would not be much out of the ordinary.

The hurricane hunters were amazed by this reaction. Use of names for storms was not new. For a great many years the worst of the world's storms have been given names, some before they struck with full force, but mostly afterward. Many were named after cities, towns or islands that were devastated. Others had gotten their names from some unusual weather that came with them or from ships that were sunk or damaged. One of them, as already has been related, was named "Kappler's Hurricane" after a weather officer named Kappler who discovered it.

During the latter part of the nineteenth century, a New Englander, Sidney Perley, collected all the available records of storms and other disasters, together with strange phenomena in New England, starting with a big hurricane in 1635, when there were only a few settlers, and continuing down to 1890. His book, *Historic Storms of New England,* was published in Salem, in 1891. He listed floods, earthquakes, dark and yellow days, big meteors, eclipses, avalanches, droughts, great gales, tornadoes, hurricanes, and storms of hail and heavy snow. Prominent among them were the "Long Storm" of 1798, the "September Gale" of 1815, and the "Lighthouse Storm" of 1851.

The "Long Storm," as the name suggests, was of long duration. It began on the seventeenth of November and continued with terrific gales and heavy snow until late on the twenty-first. This violent weather was unprecedented so early in the winter. From Perley's account it seems that the center of the storm crossed Cape Cod. A great many vessels were lost and there was much suffering among the people.

The "September Gale" of 1815 became famous because of a poem written in later years by Oliver Wendell Holmes, who was six years old at the time of the big gale. Holmes remembered and lamented the loss of his favorite pair of breeches, in part as follows:

"It chanced to be our washing day,
And all our things were drying;
The storm came roaring through the lines,
And set them all a flying;
I saw the sheets and petticoats
Go riding off like witches;
I lost, ah! bitterly I wept,—
I lost my Sunday breeches."

Holmes entitled the poem *The September Gale* and so this became the name of the storm. Actually, it was a hurricane quite like those that struck New England in 1938, 1944 and 1954. Years afterward, a New Haven man named Noyes Darling became interested in the storm of 1815 and traced its course by a collection of newspaper accounts from many places and by the logs of ships which had been in the western Atlantic when the hurricane passed. In 1842, he plotted all this information on a map and was able to figure its course. This was rather remarkable, for a study since that time shows that the tracks of hurricanes which do great damage in New England must adhere closely to one path—far enough eastward to clear the land areas as they go northward and far enough westward so that they do not go out into the ocean before they reach the latitude of Nantucket. Those which strike shore to the southward may reach New England but passage over land causes them to lose much of their fury on the way. Darling's plotted path was correct according to experiences since that time.

The "Lighthouse Storm" of 1851 commenced in the District of Columbia on Sunday, April 13, reached New York on Monday morning, and during the day struck New England. It came at the time of the full moon and so the storm-driven waters joined with the high tides, and the sea, rising over the wharves at Dorchester, Massachusetts, came into the streets to a greater height than had ever been known before. All around the coasts of Massachusetts and New Hampshire there was much property damage. The event which gave the storm its name was the destruction of the lighthouse on Minot's ledge, at Cohasset, Massachusetts. It was wrecked and swept away. At four o'clock the morning after the storm some of the wreckage was found strewn along the beach. Two young men, assistant light keepers, were killed. Since this was a very dangerous rock and many vessels had been lost there, a new lighthouse was erected at the same point soon afterward.

One of the most noted storms of the nineteenth century was "Saxby's Gale," which caused a great amount of destruction in New Brunswick on October 4, 1869. The amazing fact was that this storm was predicted nearly a year before by a Lieutenant Saxby of the British Navy. In November, 1868, he wrote to the newspapers in London, predicting that the earth would be visited by a storm of unusual violence attended by an extraordinary rise of tide at seven o'clock on the morning of October 5, 1869.

Saxby wrote the following explanation of his forecast to the newspaper:

"I now beg to state with regard to 1869 at 7 A.M. October 5th, the Moon will be at the part of her orbit which is nearest the Earth. Her attraction will be therefore at its maximum force. At noon of the same day the Moon will be on the Earth's equator, a circumstance which never occurs without marked atmospheric disturbance, and at 2 P.M. of the same day lines drawn from the Earth's centre would cut the Sun and Moon in the same arc of right ascension (the Moon's attraction and the Sun's attraction will therefore be acting in the same direction); in other words, the new moon will be on the Earth's equator when in perigee, and nothing more threatening can, I say, occur without miracle. The earth it is true will not be in perihelion by some sixteen or seventeen seconds of semidiameter.

"With your permission I will during September next (1869) for the safety of mariners briefly remind your readers of this warning. In the meantime there will be time for the repair of unsafe sea walls and for the circulation of this notice throughout the world."

It seems that Saxby had made other similar forecasts. Commenting on one of his predictions, a London newspaper, the *Standard,* said:

"Saxby claims to have been successful in some of his predictions, and he may prove either lucky or clever on the present occasion. As the astronomical effect will operate over the entire globe, it is exceedingly likely there will be a gale of wind and a flood somewhere."

The extraordinary fact is that a citizen of Halifax, Nova Scotia, disturbed by Saxby's prediction for October 5, 1869, wrote to the local newspaper the week before:

"I believe that a heavy gale will be encountered here on Tuesday next 5th October beginning perhaps on Monday night or possibly deferred as late as Tuesday night, but between these two periods it seems inevitable. At its greatest force the direction of the wind should be southwest, having commenced at or near south.

"Should Monday the 4th be a warm day for the season an additional guarantee of the coming storm will be given. Roughly speaking the warmer it may be on the 4th, the more violent will be the succeeding storm. Apart from the theory of the Moon's attraction, as applied to Meteorology—which is disbelieved by many, the experience of any careful observer teaches him to look for a storm at next new moon, and the state of the atmosphere, and consequent weather lately appears to be leading directly not only to this blow next week, but to a succession of gales during next month."

Actually the fourth began as a warm day in New Brunswick and later in the day the storm became violent, as predicted by the Halifax citizen, named Frederick Allison.

There were high tide and heavy rain at Halifax but the weather in general was a disappointment, for the citizens, after seeing the warning in the newspaper, had made many preparations about the wharves, moving goods to higher floors in warehouses, and anchoring boats out in the stream or securing them with lines in all directions.

Near by in New Brunswick, however, the storm on October 4 was severe. The gale rose to hurricane strength between 8:00 and 9:00 P.M. The tide at St. John was above any preceding mark. Vessels broke away from their moorings and some were badly damaged. Buildings were flooded and in St. John and other cities and towns in the area, buildings were demolished or unroofed, tracts of forest trees were uprooted, and cattle were drowned in great numbers.

All of this was rather remarkable as the storm reached its height at about 9:00 P.M. on October 4th, which was actually after midnight by London time and therefore on October 5th. Regardless of these circumstances, this is an instance of a storm that had a name—"Saxby's Gale"—long before it occurred and for years afterward. Some weathermen thought that it was of tropical origin and had been a hurricane in lower latitudes, but if so, it came overland in its final days, for it was felt at Washington, Baltimore, Philadelphia, and in parts of New England on the third and early on the fourth, with heavy rains and gales in many localities.

A few hurricanes have been named for the peculiar paths they followed. One that was very unusual was the "Loop Hurricane" of October, 1910. It was an intense storm that passed over western Cuba, after which its center described a small loop over the waters between Cuba and Southern Florida. When it finally crossed the coast of western Florida, it caused tides so high that many people had to climb trees to keep from drowning. The "Yankee Hurricane" was so named by the Mayor of Miami. It was first observed to the east of Bermuda in late October, 1935, moving westward. On approaching the coast of the Carolinas, it took an extraordinary course, almost opposite to the normal track at that season, and went southwestward to southern Florida, with its calm center over Miami on the fourth of November. In the same year, another unusual storm known as the "Hairpin Hurricane" started in the western Caribbean, moved northeastward to Cuba, and then turned sharply southwestward to Honduras, describing a track shaped like a hairpin. It caused one of the worst disasters of that region. Loss of life exceeded two thousand.

Examples of storms named after ships are "Racer's Storm" in 1837, named after a British sloop of war which was caught in its hurricane winds in the Yucatan Channel. Another one of great violence was called "Antje's Hurricane," because it dismasted a schooner of that name in the Atlantic in 1842.

In Puerto Rico, a hurricane may be given the name of the saint whose feast is celebrated on the day on which it strikes the island. The most famous are: Santa Ana, July 26, 1825; Los Angeles, August 2, 1837; Santa Elena, August 18, 1851; San Narcisco, October 29, 1867; San Felipe, September 13, 1876; San Ciriaco, August 8, 1899; and the second San Felipe, September 13, 1928.

Doubtless the worst hurricane during the twentieth century was the one in 1928, "San Felipe." It caused damage estimated at fifty million dollars in Puerto Rico, and later struck Florida, causing losses estimated at twenty-five million dollars. Puerto Rico lost three hundred lives, Florida nearly two thousand.

One of the well-known storms of the West Indies was the "Padre Ruiz Hurricane," which was named after a priest whose funeral services were being held in the church at Santa Barbara, Santo Domingo, on September 23, 1837, when the hurricane struck the island, causing an appalling loss of life and property destruction.

Before the end of the nineteenth century, a weatherman in Australia named Clement Wragge had begun giving girls' names to tropical storms. Down in that part of the Southern Hemisphere, hurricanes are called willy-willies. They come from the tropics on a southwest course and turn to the south and southeast on approaching or passing Australia. Their winds spiral inward around the center in a clockwise direction—the opposite of the turning motion of our hurricanes.

Wragge was the government meteorologist in Queensland, and later ran a weather bureau of his own in Brisbane. A tall, thin, bewhiskered man who stammered, he was known all over Australia as a lecturer on weather and similar subjects. Australians of that time said that, as likely as not, when due to talk about big winds, he would arrive at the lecture hall with "too many sheets out" and fail to keep on his feet during the lecture. Though his name was Clement, he was better known in Australia as "Inclement."

Storms which did not come from the tropics were called by men's names. Generally, Wragge called them after politicians who had earned his disfavor, but for some reason he used girls' names for the willy-willies. As an illustration for his weather journal called "WRAGGE," he had a weather map for February 2, 1898, with a willy-willy named "Eline." He predicted nasty weather from a disturbance named "Hackenbush."

E. B. Buxton, a meteorologist for Pan American Airways, went to the South Pacific in the late thirties and, after hearing about Wragge and his names for willy-willies, adopted the idea for his charts. He recalled particularly using the name "Chloe" for hurricanes.

With few exceptions, the hurricanes of the twentieth century went unnamed in the United States until 1951, although some were referred to in terms of place and date; for instance, the "New England Hurricane of 1938." Unofficially, a few had names of people. In 1949, while President Truman was in Miami addressing the Veterans of Foreign Wars, the first hurricane of the season was called "Harry," and a little later a bigger one which the newsmen said had greater authority struck southern Florida and it was called "Hurricane Bess."

In sending out advices and warnings of West Indian

storms, it was not considered necessary to have names, as it was seldom that more than one was in existence at the same time. In 1944, when aircraft reconnaissance began, it became customary to get reports by radio-telephone and voice was used increasingly in other ways by the hurricane hunters. But this gave no particular trouble until September, 1950, when there were three hurricanes in progress at the same time.

Two were in the Atlantic, one north of Bermuda and the other north of Puerto Rico. The third appeared in the eastern Gulf of Mexico. When aircraft were dispatched into these storms and began reporting, there was increasing confusion. Other communications and public advices became mixed and there was much uncertainty as to which storm was meant. Use of letters of the alphabet to identify them was no help, for letters B, C, D, E, and G sound much alike by radio-telephone; also A, J, and H. Numbers were no better because weather reports are sent by numbers and the advisories issued on each storm are numbered, so that the number 3 could be the number of the storm, the number of the advice, an element of the weather, the hour, etc.

The agencies involved in weather and communications in connection with hurricanes met early in 1951 and decided to identify storms by the phonetic alphabet, which gave Able for A, Baker for B, Charlie for C, etc., in accordance with the following table:

Able	Jig	Sugar
Baker	King	Tare
Charlie	Love	Uncle
Dog	Mike	Victor
Easy	Nan	William
Fox	Oboe	Xray
George	Peter	Yoke

How	Queen	Zebra
Item	Roger	

In the 1951 season, this worked very well in the communications and the public began to speak of hurricanes by these names. At the start of the 1952 season, the agencies began to use the same list of names, starting with Able for the first storm, but soon ran into difficulty. A new international alphabet had been introduced as follows:

Alfa	Joliet	Sierra
Bravo	Kilo	Tango
Coca	Lima	Union
Delta	Metro	Victor
Echo	Nectar	Whiskey
Foxtrot	Oscar	Extra
Golf	Papa	Yankee
Hotel	Quebec	Zulu
India	Romeo	

Some of the agencies had begun using the new alphabet in their communications, while others stuck to the old one. So the third storm of the season was "Charlie" part of the time and the rest of the time some wanted to call it "Coca." At the end of the season there was no agreement as to which phonetic alphabet should be used and there was criticism for having continued an alphabet which was obsolete internationally.

After a long discussion, military members of the conference suggested adoption of girls' names, which had been used successfully for typhoons in the Pacific for several years. Just how this practice originated is not known, but it was thought by some persons to have come from the book *Storm,* by George R. Stewart, which was published in 1941. In this book a fictitious Pacific storm is traced to the United States and its effects on the people are narrated in the style of a novel. A young weatherman at San Francisco, according to the story, called the storm Maria. Also there was Wragge's use of girls' names for willy-willies in Australia and Pan American Airway's practice in connection with hurricanes as early as 1938. At any rate, with these Pacific precedents, the weathermen and hurricane hunters adopted the following list for 1953 for hurricanes in the Atlantic, Caribbean and Gulf of Mexico:

Alice	Irene	Queen
Barbara	Jill	Rachel
Carol	Katherine	Susie
Dolly	Lucy	Tina
Edna	Mabel	Una
Florence	Norma	Vicky
Gilda	Orpha	Wallis
Hazel	Patsy	

This list worked perfectly in 1953; the public was pleased; the communicators were happy about it; the newspapers thought it was colorful; and use of the same names began to spread in Canada and some of the countries to the southward. The same list was adopted with enthusiasm for the 1954 season.

In 1954, Alice and Barbara were minor hurricanes in the Gulf of Mexico, although Alice broke up in tremendous rains in the upper watershed of the Rio Grande, after moving inland over Mexico. There were floods which broke records for all time as the water moved down the river. The third storm, Carol, started a controversy in the press and many letters were written to the editors and to the Weather Bureau, some favoring the scheme or trying to get a little fun out of it, but most of them finding objections of one kind or another. It was almost impossible to change in the middle of the season, even if the hurricane hunters had wanted to, so it was continued during 1954 and each new hurricane aroused further comment. Later Hazel came along about the middle of October, a very severe hurricane from the Caribbean Sea. It turned northward between Cuba and Haiti and caused terrible damage and much loss of life. Later it struck the coast of the Carolinas and crossed the eastern states northward to New York. Loss of life in the eastern states was variously estimated from fifty to eighty, and the damage to property, especially from falling trees, was enormous. There was another flood of complaints, this time about the name Hazel.

Before the argument was ended it threatened to be almost as stormy as some of the smaller hurricanes so named. Early in 1955 the Weather Bureau had a meeting with the Air Force, Navy and others interested in deciding the question. By that time the opinions received by mail were overwhelmingly in favor of continuing girls' names. In the meantime, there had been a surprise. A storm having some of the characteristics of a hurricane was sighted in the Caribbean Sea in January and, in the absence of a decision on names to be used in 1955, it was called Alice from the 1954 list. Later, the names for others in 1955 were decided as follows:

Brenda	Janet	Rosa
Connie	Katie	Stella
Diane	Linda	Trudy
Edith	Martha	Ursa
Flora	Nellie	Verna
Gladys	Orva	Wilma
Hilda	Peggy	Xenia
Ione	Queena	Yvonne
		Zelda

17. THE GEARS AND GUTS OF THE GIANT

"he that wrestles with us strengthens our nerves and sharpens our skill"—Burke

All through this book we have talked about hurricane hunters. By now it is clear that the crew on the plane that goes into the storm at the risk of destruction of the craft and death to the men is not really "hunting" a hurricane. It is the exception rather than the rule when they discover a tropical storm. The first hint comes from some distant island or a ship in the gusty wind circle where the sea and the sky reveal ominous signs of trouble. Somewhere in a busy weather office a large outline map is being covered with figures and symbols. Long, curving lines across a panorama of weather take shape as the radios vibrate and the teletypewriters rattle with the international language of weathermen—the most co-operative people in the world's family of nations.

Hurricane hunting is done on these maps. Day after day, without any fanfare, the weathermen search the reports spread across this almost boundless region where hundreds of tropical storms could be in progress if nature chose to operate in such an eerie fashion. Even the experienced observers on islands and the alert officers on shipboard might not see the real implications in the weather messages they prepare. In the enormous reaches of the belt of trade winds, where the tremendous energy of the sun's heat and the irresistible force of earth rotation dictate that the winds shall blow as steady breezes from the northeast, somebody might put in his report, for example, that there was a light wind coming from the southwest. That fact alone would be enough. In season, the weathermen would know, almost with certainty, that there was a tropical storm in the area.

There are many things to watch for, in the array of elements at the surface, in the upper air, the clouds, sea swells, change of the barometer, faint earth tremors. A hint from this scattering of messages in the vast hurricane region starts the action. And the planes go out to investigate.

This is an extraordinary procedure. Looking at it as an outgrowth of the insistent demands of citizens along the coasts in the hurricane region for warnings of these storms, as the population increased and property losses mounted, it seems that the flight of planes into these monstrous winds is justified only until a safer method can be found. All other aircraft are flown out of the threatened areas, obviously because the winds are destructive to planes on the ground. The lives of men and the safety of the plane in the air should not run a risk of being sacrificed if it can be avoided. Of course, it is argued by some men that there is a possibility that a method may be discovered to control hurricanes by the use of chemicals or some other plan requiring planes to fly into the centers. And it is true, also, that for the time being at least there is certain information that can be obtained in no other way.

At the end of World War II, there was a grave requirement for more information about hurricanes. Little was known except in theory about their causes, maintenance, or the forces which determine their rate and direction of travel. Since that time, literally thousands of flights have been made into hurricanes and typhoons. Scientists have studied the detailed records of these many penetrations.

We have learned a great deal in these years but by no means enough. Herbert Riehl, a professor of meteorology at Chicago University, has examined as large quantities of the data as any man. Recently he said, "Our knowledge regarding the wind distribution within tropical storms and the dynamical laws that guide the air from the outskirts to the center of the cyclone is so deficient as to be deplorable."

From the scientific point of view, remarks of this kind are fully justified, but progress in the issuance of warnings is quite another matter. Hurricane prediction for the present and the near future is an art and not a science. Very great progress has been made in recent years in sending out timely warnings. There are figures to show the facts. At the beginning of this century, a hurricane causing ten million dollars in property damage was likely to take several hundred lives. Twenty-five years later, the average was about 160 lives. Ten years later (1936 to 1940 average) the figure had been reduced to about twenty-five and was steadily going down. After men began flying into hurricanes, the figure was reduced to four (1946 to 1950). This is astonishing, not only in showing how the warnings were improving after hunting by air got started, but also the big gains shortly before that time, especially after the hurricane teletypewriter circuit was installed around the coast in 1935. Experience in prediction, on-the-spot operation, and fast communications are vital.

In fact, the record was so good at the beginning of World

War II that most forecasters despaired of their ability to keep it up. It had consistently been below ten lives for ten million dollars' damage and one serious mistake could have raised this rate considerably for several years. For this reason, as well as many others, the forecasters were extremely grateful for the information from aircraft.

The main hope for greater savings in the future is that the solution of some of the mysteries of the hurricane will enable the forecasters to send out accurate warnings much farther in advance. In such an event, it will be possible to protect certain kinds of property and crops which are being destroyed at present. Heavy equipment can be moved and certain crops can be harvested in season, if plenty of time is available. These precautions are time-consuming and costly, and the advance warnings must be accurate in detail. And it will help to make sure that no hurricane different from its predecessors will come suddenly and catch us off guard and cause excessive loss of life. Now and then we have one which is called a "freak."

One thing we have become increasingly sure of and it will stand repetition. No two hurricanes or typhoons are alike. Scientists may find some weather element that seems to be necessary to keep the monster going, and then are frustrated to find that not all tropical storms have it. If some can do without it, maybe it is not necessary, after all. And yet all of them fit a certain direful pattern; there is nothing else that resembles these big storms of the tropics. Like the explosion of an atom bomb, with its enormous cloud recognized by everyone who sees a picture of it, the hurricane has well-known features—unlike anything else—but of such enormous extent that no one can get a bird's-eye view of the whole. Putting together what we know by radar, upper air soundings, aircraft penetrations and millions of weather observations in the low levels, we can draw a sketchy word picture. Looking down from space, we could see it as a giant octopus with a clear eye in the center of its body, arms spiraling around and into this body of violent winds around the eye—all of the monster outlined by the clouds which thrive as it feeds on heat and moisture. We feel sure of that much.

The birth of the THING has not been explained. There are plenty of times when all the ingredients are there. Nothing happens. Observation and theory flourish and swell into confusion. No scientist can say, "Everything is just right; tomorrow there will be a hurricane."

Why it moves as it does is another grim puzzle. Ordinarily, the great storm marches along with the air stream in which it is embedded, changing its path with the contours of the vast pressure areas which outline the circulation of the atmosphere, but too often it suddenly changes its mind, or whatever controls it, or shifts gears, and comes to a halt, or describes a loop or a hairpin turn. Nobody can see these queer movements ahead of time. Going out there in an airplane to look the situation over does not help in this respect. It is a vital aid in keeping track of the THING and protecting life and property, but it ends there.

Where does all the air go? When the big storm begins out there over the ocean, air starts spiraling inward and the pressure falls, showing that the total amount of air above the sea to the top of the atmosphere is lessening, even as it pours inward at the bottom. For a hundred years scientists argued that it must flow outward at the top, that at some upper level the inflow of air ceases and above that there must be a powerful reversal of the circulation. Here again we have frustration. Going up with one of the investigators, we get the facts. Strangely enough, this is one of the men who want to get into hurricanes, who come down to the coast to look, and who finally "thumb a ride" with the airmen into the big winds. A brief of his story will illustrate.

This story begins with the big Gulf hurricane of 1919. It came from the Atlantic east of the Windward Islands, moved slowly to the northward of Puerto Rico and Haiti and thence to the central Bahamas, a fairly large storm threatening the Atlantic seaboard. Then it took an unusual path, generally westward, with increasing fury. It was a powerful storm as its central winds ravaged the Florida Keys and took a westward course across the Gulf. It happened shortly after World War I and there was little shipping in the Gulf. The slow-moving hurricane, now a full-fledged tropical giant, dawdled in the Gulf and was lost; that is, lost as much as a monster of its dimensions can be, but its winds were felt all around the Gulf Coast and its waves pounded the beaches as it spent four days out there without disclosing the location or motion of its calm center.

Warnings flew all around the coast and the week dragged to an end with the people extremely tired of worrying about it and the weathermen worn out with continuous duty. Saturday night came and the center seemed to be no nearer one part of the coast than another. Late at night, an annoying thing happened. It was customary in those days for the forecaster, in sending a series of messages from Washington, to stop them at midnight and begin again early the next morning. It was the rule that no reports came in between midnight and dawn. The clerk sending the last message added "Good Night," to let the coastal offices know that there would be no more until morning.

In this case, the forecaster ended his advisory with a notice putting all Gulf offices on the alert and the clerk added "Good Night." And so the offices received a message ending with these words: "All observers will remain on the alert during the night. In case the barometer begins to fall and the wind rises, Good Night." This created a furor in coastal cities on the West Gulf and it was several weeks before the criticism subsided. By Sunday morning, however, the gusty wind had not risen much and there was no great fall in the barometer, so the weathermen had no answer at daybreak. Soon afterward, however, the weather deteriorated rapidly at Corpus Christi, and hurricane warnings went up as big Gulf waves pounded over the outlying islands into Corpus Christi Bay and the wind began screaming in the palms.

Around noon the worst of it struck the city. The tide mounted higher than in any previous storm of record, except in the terrible Galveston hurricane of 1900. Much of Corpus Christi was on a high bluff above the main business section, but the latter and the shore section to the north were low. It was after church and time to sit down to Sunday dinner when the final rise of the water begain to overwhelm everything. The police, sent out by the Weather Bureau, were knocking on people's doors and telling them to get out and run for high ground. But these low sections had survived a big, fast-moving hurricane three years before, without nearly so high a tide, and most people thanked the police but determined to stay and eat. This decision was fatal in the North Beach section. The road was cut off and nearly two hundred were drowned.

Down on Chaparral Street lived a man named Clyde Simpson, with his wife and seven-year-old son Robert. The boy's uncle and grandmother were there also. They were about to sit down to a big platter of chicken, and the boy had his eye on a pile of freshly fried doughnuts. They had been out standing with other nervous people to look at the great waves roaring across the beach, but after a little the storm waters had forced them back and covered the streets. Now the water was rising fast. Several houses had come up off their foundations. A large frame residence on the opposite side of the street floated across, and, while they held their breath, missed them by a few feet, struck the house next door, and both collapsed. The elder Simpson said it was time to get out, dinner or no dinner.

The family went through the back yard, the nearest route to higher ground. The boy's mother put the dinner in a large paper sack and held it above her head as she struggled through the water. The father carried the seven-year-old on his back and brought up the rear, swimming a little as the water continued to rise. The grandmother, an invalid strapped in a wheel-chair, was pushed and floated ahead by the uncle. The boy worried as his mother got tired and let the paper sack hang lower and lower. Finally it hit the water and the chicken and doughnuts sank or floated away. That scene was etched in Robert's memory, along with the battering of the winds and the tremendous rise of the waters over the stricken city. The family survived.

Looking out of the windows of the courthouse on the edge of the bluff above the business section, the boy watched others struggling toward higher ground. Afterward the family returned to their house, smeared with oil and tar and by dirty water, floors covered with sand, mud, and debris. Robert saw death on every hand—dead dogs, birds, cats, rodents, and one neighbor who failed to get out.

In 1933, when one of the hurricanes of that year crossed the Gulf and threatened the lower Texas Coast, much like the big one in 1919, a young fellow drove all the way from Dallas to have a look at it. He was Robert Simpson. He never got it out of his mind. Finally, he joined the Weather Bureau, worked at hurricane forecasting offices and in 1945 "thumbed" his first ride into a hurricane. After that his enthusiasm and persistence annoyed some of the older weathermen and bothered members of the air crews who flew the big storms both in the Atlantic and Pacific.

Simpson made up his mind that he would use every opportunity to find out how the big storms were organized and what they were geared to in their movements, regular and irregular—the gears and guts of the THING. When Milt Sosin lurched into the center of the big storm in 1947 in a B-17 and looked up to see a B-29 high in the eye of the same hurricane, Simpson was up there with the men from Bermuda, trying to find out what steered the monster. And on this flight, with a B-29, they expected to come out on top at twenty-eight to thirty thousand feet, according to the theorists and the textbooks, but they broke out just below forty thousand, still one hundred miles from the center. From there the high cloud sheet should have sloped downward to the center, if they were to believe the accepted doctrine of circulation in the top of the hurricane. But they were shocked and chagrined to find that the high cloud sheet—the cirrostratus—sloped sharply upward in front of them, rising far above the extreme upper operational ceiling of the B-29.

And so the superfortress turned toward the center and rocketed into the high cloud deck with misgivings on the part of Pilot Eastburn and Simpson. The latter reported:

"Through this fog in which we were traveling at 250 miles an hour there loomed from time to time ghost-like structures rising like huge white marble monuments through the cirrostratus fog. Actually these were shafts of supercooled water which rose vertically and passed out of sight overhead as we viewed them from close at hand. Each time we passed through one of these shafts the leading edge of the wing accumulated an amazing extra coating of rime ice. This kind of icing would have been easy to shake off if the plane had been fitted with standard de-icing equipment. But it was not. We were so close to the center of the storm by the time the icing was discovered that the shafts were too numerous to avoid.

"Pilot Eastburn punched me and pointed to the indicated airspeed gage. It stood at 166. 'At this elevation this plane stalls out at 163,' Eastburn said, 'and in this thin air there is no recovery from a stall.' He continued, 'We have got to get out of here fast!' I nodded agreement, feeling a bit sheepish about the whole thing. After all, hadn't Vincent Schaefer, of General Electric, just a few months earlier demonstrated in the laboratory that water vapor could be cooled to a temperature of —39° before freezing set in? But in the turbulent circulation of a hurricane—this was fantastic! Unbelievable! But there certainly was no guesswork about that six or eight inches of rime ice on the leading edge of the wing!

"We got out of there all right, and fast, but we had to do it in a long straight glide; the plane was simply too loaded with ice and too near stall-out to risk the slightest banking action."

After all, the atmosphere is a mixture of gases and it obeys the laws of gases. If the scientists assume that the big storm has a certain structure and a certain circulation of air in its colossal bulk, there are definite conclusions to be drawn concerning the physics of this giant process in the tropical atmosphere. But if it turns out that the assumptions about the structure and circulation are wrong, the conclusions of the physicists may be exactly opposite to the truth. The results of years of study, calculation and discussion seem to be overthrown in one moment as a superfortress plunges into a vital section and the crew sees things that ought not to be there!

Most important in the 1947 storm was the fact that conditions at a height just below forty thousand feet were such as to go with a circulation *against* the hands of a clock at maybe 130 miles an hour. The plane going in that direction had a tail wind of ninety miles an hour. And yet, the students of hurricanes during the past century were sure that at some height well below that level the winds blew outward in a direction *with* the hands of a clock. In agreement with this conclusion, most of the scientists had made up their minds in recent years that the circulation in the lower part of these storms usually disappears at twenty to thirty thousand feet. And so, if we are to account for the removal of air in this great space extending down to the sea surface, it must have been done well above forty thousand feet in this case. And up at this height the air is so thin that it is almost inconceivable that it could blow hard enough to account for air removal in the average hurricane. On the other hand, this was a mature storm and it may be that at this stage no air was actually being removed from the system and that the gigantic circulation of the full-grown monster is self-contained.

While it would be extremely interesting to understand the magic by which nature so slyly removes the air from the hurricane under our very noses, the practical question is whether or not its escape at the top is geared in any way to the forward motion of the main body of the storm. The answer to the first question may give the answer to the second, and possibly also to the third question: what causes a hurricane to increase in intensity—to deepen, as the weatherman says, having reference to the fall of pressure in the center? He thinks of it as a hole in the atmosphere.

This 1947 hurricane illustrates the great difficulty of finding answers to our questions. But in any case, this was just one storm and all of them are different in one way or another.

But to go back to the story of the guest rider from the Weather Bureau, Robert Simpson, the story is not complete without a brief account of the flight into Typhoon Marge. It raised its ugly head in the Pacific in August, 1951, and on the thirteenth had passed Guam, a storm not well developed but of evil appearance, showing signs of growth. That evening Simpson arrived from Honolulu, where he was in charge of the Weather Bureau office. He accepted an invitation from the Air Force to visit Marge and on August 14, six hours after he alighted from Honolulu, was airborne in a B-29 and on the way.

In a few hours Marge had grown into a colossus. It was nearly one thousand miles in diameter, with winds exceeding one hundred miles an hour in an area more than two hundred fifty miles in diameter. When the hurricane hunters entered the center and measured the pressure, it proved to be one of the deepest on record—26.45 inches at the lowest point. From plane level, the eye was perfectly clear above, forty miles in diameter and circular. The massive cloud walls around the eye rose on all sides to thirty-five thousand feet, like a giant coliseum. The west wall was almost vertical, with corrugations that suggested the galleries of a gigantic opera house.

In the center, below the plane, they saw a mound of clouds rising to about eight thousand feet, an unusual feature, but one that has been observed in other tropical storms. The crew spent fourteen and a half hours in the central region of this huge typhoon, getting data at levels from five hundred feet up to twenty thousand. Down in the lower levels, they found a horizontal vortex roughly five thousand feet in diameter, extending from the cloud wall of the eye like a tornado funnel, in which they encountered very severe turbulence. Another collection of data was added to the growing accumulation and with it the notes of unusual phenomena observed. Since that time Simpson has flown several hurricanes in the Atlantic.

Now it is abundantly clear that the hurricane hunters are looking for many important facts aside from the location of the tropical storm and a measure of its violence. There are many questions unanswered. Here in the warm, moist winds that blow endlessly across deep tropical waters there are mysteries that have challenged man for centuries. Turning to their advantage every discovery that science has pointed in their direction, the hurricane hunters have cheated the big storms of the West Indies of a very large share of their toll of human life. In struggling to solve the remainder of the problem, they have two virtues that will ultimately bring success—ingenuity and persistence. They push on tirelessly in several hopeful directions.

The Navy has taken advantage of the strange fact that when a tropical storm comes along it literally shakes the earth. There are little tremors like earthquakes but very much smaller. The Greek word for earthquake is *seismos* and by putting *micro* in front, meaning very small, we have the word *microseism.* And so, the storm-caused little tremors are called microseisms or slight earthquakes. The instrument which registers these tremors is called a seismograph. When the earth moves, even a very little, a body on the earth tends to hold its position and the earth moves under it. In a small earthquake, a chair will move across the floor. This kind of motion can be registered by instruments.

In 1944 the Navy installed seismographs and began keeping records of the slight tremors caused by hurricanes and typhoons. These studies have shown that a tropical storm at a distance produces a small tremor which becomes stronger as the storm center gets nearer. No one knows exactly how the storm shakes the earth and causes the tremors. There are some strange things about this. It seems that these microseisms are carried along in the earth until they come to the border of a great geological block and then do not pass readily into the next block. So there are places in the Caribbean where the tremors weaken as they come to a different earth block and this interferes with the indications picked up by the instruments. The fact is that microseisms give signs of the existence of a tropical storm and sometimes serve to alert the storm hunters, but they are by no means good enough to replace the use of planes in tracking them. But the studies of microseisms are being continued.

For many years static on the radio, better known as atmospherics or just "sferics," has been used in the endeavor to locate or keep track of storms. At first the Navy tried it on West Indian hurricanes. The instruments used will find the direction from which the sferics come when they are received in a special tube. In more recent years, the Air Force has used this scheme. It works to advantage in finding thunderstorms, but tropical storms are so big and the sferics are not found in any regular pattern around the central region. After years of trial, it has been concluded that this scheme is not good enough to replace other methods.

Of all the methods of this kind, radar is by far the best. But as the radar stations on shore and the radar equipment on aircraft have increased in numbers and have been improved to reach greater distances, some new troubles have arisen. For many years the hurricane hunters took it for granted that a hurricane has a clear-cut center which moves smoothly along a path that is a straight line or a broad curve, but in a few cases is a loop or a sharp turn. In other words, the center does not change size and shape or wiggle around. In the past, when an observer on a ship or on a plane reported a center of an odd shape or had it off the smooth path the hunters were plotting, they said the observer had made an error.

Now as the hunters have begun watching hurricane centers close by on the radar, they see them changing shape and wiggling around. In fact, as stated in a few cases in earlier chapters, they have seen false eyes and have been confused by them until the true eye came into view on the radar scope. If the true eye describes a wiggly path and changes size, the hunters can draw the wrong conclusions about its direction of motion unless they wait a while to see if it comes back to the old path. The hurricane is a little like an eddy or whirl in water running out of the bottom of a bowl. It is a violent boiling eddy that twists and changes shape, and in a substance as thin as the atmosphere these motions are not steady to such a degree that the observer can reach a quick decision. At any rate, it is now apparent that the observers on ships and aircraft did not make as many errors as was thought several years ago.

There is another aspect that must be kept in mind. Radar shows areas where rain is falling around the center of a hurricane and so the center, having no rain, stands out as an open space on the radar scope. This is very good if the storm has rain all around the center, but some of them have very little rain on the southwest side, and in some cases there is none to return an echo to the radar. In such a case, there is only one side to the storm echo and the location of the center is not revealed. Of course, these facts are known to the experienced radar men, but they should be known to everybody interested in hurricane reports; otherwise they are likely to expect too much accuracy from observations of this kind.

For these and other reasons, the man on the aircraft has a very great advantage in daylight, for he can see clouds of all kinds, measure the winds and, by moving through the storm area at the speed of the modern plane, he can see a large part of it in a short time. To find a substitute for aircraft reconnaissance is going to be extremely difficult. But at night the situation is quite different. The airman is unable to see much without radar, except on a moonlight night and that is not very good.

One suggestion that has been put forward by a number of different people in recent years is that a balloon be flown in the calm center and followed by radar or radio, thus keeping track of the storm's motion. It is possible, of course, to fix a small rubber balloon (perhaps eight to ten feet in diameter) so that it will remain at the same height for a fairly long time. By one method the rubber balloon is partly filled with helium and covered loosely with nylon. The balloon expands as it rises, becoming less dense as the atmosphere gets thinner. It continues to rise until it fills the nylon cover and cannot expand further. After that, its density becomes the same as the air at some level previously chosen, and from there it drifts along without rising or descending.

It is the idea that the obliging balloon would drift here and there in the vagrant breezes of the eye, but when it came to the edge of the powerful wind currents around the outside of the eye it would be guided back in. No experiment has been carried out to prove that this would happen but such trials have been scheduled and will be made at the first opportunity. There is one difficulty. The question is how to get an inflated balloon into the center and release it under proper conditions. One of the men who has worked on a scheme of this kind is Captain Bielinski, the Air Force officer who broke his hundred-dollar watch in a typhoon and solemnly swore he would find an easier way to do it. He calls his device "Typhoon Homer." He has worked on it for four years, spending much of his own time and money.

There are reasons to believe that, after a few experiments, a height could be found where the balloon would stay in the eye. So far as we know, birds trapped in the center are held there. After battling hurricane winds, they are so exhausted on getting into the center that they could not remain there if the wind circulation tended to suck them out into the surrounding gales.

Bielinski concluded that the balloon could not be thrown out from a plane in even a partially inflated condition. The blast of air on leaving the aircraft would destroy it or put it out of commission. So he has an uninflated balloon and bottles of gas, a small radio transmitter, and a float, all attached to a parachute.

The bottles and radio would be thrown out, the parachute would open, and the gas would go through a tube from the bottles into the balloon. The float, with a long line to the balloon, would rest on the water and provide an anchor for the apparatus. The radio would send signals every hour, the operators on shore would figure its location by direction finding, and there would need to be no further aircraft flights into that storm. The device, according to Bielinski, would continue to operate for seven days.

Robert Simpson and others have had similar ideas, some favoring a device that could be followed by radar, but Simpson prefers the radio transmitter. To find out how the air circulation in the calm center would affect the balloon, he planned experimental flights in hurricanes to release a chaff made of a substance that could be followed by radar. He tried it in 1953 and again in 1954, but something happened in each case to prevent the experiment from being carried out. In one case, for example, nearly everything was in readiness for an experimental flight to take off when Edward Murrow of CBS arrived in Bermuda with his crew and apparatus to put Hurricane Edna on television, and Simpson was moved to the back of the plane. He and all others connected with it, including Major Lloyd Starret, who had been brought in from Tinker Air Force Base to work with Simpson, were glad to make way for a public service program. But this shows one of the reasons why developments of this kind, which depend on opportunities in only a few hurricanes a year, take a discouragingly long time. There was no chance to test Bielinski's device, or any other, for that matter. There have been laboratory experiments also on a device to deflect the air streams around the bomb bay of the aircraft so that a partially inflated balloon could be safely released in the eye of a storm.

These devices are mentioned here to show the trend of thought. Something similar to this may eventually serve to replace a large share of the hazardous aircraft flights, but even if the center is satisfactorily located in such a manner, much useful information on the size of the storm, the force of its winds, and other data will be determined in many cases only by aerial reconnaissance. With this in mind, both the Air Force and Navy are substituting bigger and better aircraft for this purpose.

The old B-29 Superfortress is being "put out to pasture," as they say in the Air Force. The higher, faster, and farther flying Boeing B-50's are replacing them, not only in hurricane reconnaissance but in the daily flying of weather routes to help fill in the blank spaces on the world's weather charts. The B-50's will go ten thousand feet higher than the B-29's. Another advantage that appeals to the hurricane hunters who fly on these missions is the electric oven, standard equipment on the B-50, which will furnish hot meals at favorable times on the route, instead of sandwiches and thermos coffee. The Navy, not to be outdone, is coming out with the Super Constellation, which is being modified for hurricane reconnaissance to replace the P2V Lockheed Neptune recently used.

As each new season comes, the hunters are wiser and better equipped. The battle with the hurricane is joined. It is something to worry about, like war and the H-bomb. At the end of the 1954 season, the executives of the big insurance companies were in conference with grave faces. Property damage from Carol, Edna and Hazel had mounted upward to around a billion dollars. Reports had been circulated to the effect that the slow warming of the earth in the present century is bringing more hurricanes with greater violence and paths shifting northward to devastate areas with greater populations. There was speculation about the effects of A-bombs and H-bombs on hurricanes.

All this trouble comes from water vapor in the atmosphere. Without it, the earth would be a beautiful place but useless to man. Even over the tropical oceans it rarely exceeds five per cent of the bulk of the air. In other regions, it is much less. But it is this vapor, constantly moving from the oceans into the air and spreading around the world, that builds the stormy lower layer of our atmosphere—the troposphere—where clouds and storms, snow and ice and torrential rain, thunderstorms, hurricanes and tornadoes thrive in season. Such tremendous energy is needed to carry billions of tons of moisture from the oceans to the thirsty land that all of these rain and storm processes are maintained on the borderline of violence.

Here at the bottom of the atmosphere the vapor absorbs the heat radiated from the sun. There is a swift drop in temperature as we go aloft. Moist air pushed upward becomes cooled and ice crystals, water droplets, snowflakes, are squeezed out. Clouds form, beautiful in the sunset, gloomy on a winter day, threatening as the summer thunderstorm shows on the horizon, fearsome as the winter blizzard takes command of the plains and valleys. Here is water vapor coming to the end of a long journey from the surfaces of distant seas. From here it goes to the land and begins another long journey, in the rivers and back to the oceans. But on the way to us, violence may be one of the principal ingredients. We can't live without it and we have trouble living with it.

When this lush flow of water vapor from the tropical ocean to the atmosphere becomes geared in some special manner to swiftly-moving air from other regions, the process seems to get out of nature's hands. Upward motion begins on a grand scale. Converging streams of air are twisted by the spinning of the earth on its axis. And just as men begin to see the picture, nature draws a veil by the condensation of water vapor. Under this darkening canopy, violence grows with startling swiftness. The water vapor that drew the curtain now releases energy alongside of which the A-bomb shrinks to insignificance.

Far below the sea surface, the solid earth trembles. Avalanches of water are torn from the ocean and hurled down the slopes of the gale. A colossal darkening storm begins to move across the ocean. It sucks inward the hot, moist lower atmosphere and brings it along with it, using the vapor to feed its monstrous, seething caldron. Down here at the surface of the earth, its winds are warm and humid. Its tentacles —octopus-like arms—reach out with gale-driven torrents of rain and begin picking everything to pieces. After hours that seem like days, the central fury of the earth-blasting storm begins its devastation of man's possessions.

And as it has proved to be unquestionably true that no two hurricanes are exactly alike, so it is evident now that the same hurricane is subject to massive changes from day to day. It has a life history. Like the caterpillar that is transformed into the cocoon and then into the butterfly, the tropical storm goes through definite stages. The problems involved for the hurricane hunters in each of these distinct stages demand separate solutions. Like a living thing, the monster has infancy, youth, middle age and decline.

In infancy, its malevolent forces are directed vigorously

toward the mysterious removal of large quantities of air from above its gale-swept domain. The excessive heat and moisture of its birthplace yield far more energy than is needed to keep its mighty low-level winds in motion.

In youth, it is extremely violent and the removal of air brings exceedingly low pressure into its center. Its outer parts become ominously visible through the condensation of moisture on a grand scale, cloaking its internal mechanism. Its destructive forces spread. In this stage, the removal of air in upper regions continues in excess of the inflow at the bottom in proportion to the horizontal expansion of the system.

In middle age, its violent forces are directed toward maintenance of the colossal wind system. The total energy it can derive from heat and moisture no longer produces an outflow above in excess of the inflow of air at the bottom. It expands in the vertical and its visible parts push against the stratosphere. As it moves farther away from its birthplace and the available energy begins to decline, it dies. For a few days nature's processes for the transport of moisture from the oceans to the thirsty continents have run amuck. Life and property suffered while torrential rains fell.

So it is clear that in life the monster thrives on heat and water vapor. Down at sea level it is a warm phenomenon. Only the heated air of the tropical regions can hold enough moisture to feed the giant.

But up above, the full-grown hurricane is not a warm storm. Hunters perspire at low levels but not in the top of the storm. There are icy corridors through currents of air robbed of their heat by the monster below. Pillars of supercooled water push upward into the thin atmosphere. Snow flies with the shuddering winds at the top of the troposphere. It is colder up here above the tropics than it is above the poles. The fingers of the gale tremble with the cold and seem to make gestures in defiance of the sun shining through the stratosphere. Water vapor in great quantities has been carried high in the atmosphere and nature seems powerless to bring equilibrium until land or cold water at the earth's surface below shuts off the abundant supply of energy. And when it does, the monster dies as it was born, hidden behind a veil produced by lingering cloud masses derived from the vapor that gave it life.

In the last few years, men have had the courage to fly into these monsters. Some day, when other methods are used, people will look back in amazement at these brave events. Here they can see how it happened, how it was done, and feel admiration for the men who did it—the hurricane hunters.

IVAN RAY TANNEHILL

was born in Ohio, where he obtained both his degrees in science at Denison University. While a boy in his early teens, he became intensely interested in birds, stars and the weather. After college, he joined the Weather Bureau in Texas and a year later went through a vicious hurricane at Galveston.

This experience led Dr. Tannehill to study hurricanes for the next forty years. Twenty years ago he became chief of the marine division of the U. S. Weather Bureau, then he was chief of all the Bureau's forecasting and reporting and finally was assistant chief of the Bureau, in charge of all its technical operations.

Dr. Tannehill is the author of several authoritative books on the weather, including a world-recognized classic, *Hurricanes; Their Nature and History,* now in its eighth edition. He has represented the United States at many world conferences on weather and served several years as president of the international commission on weather information. Citations, medals, awards and commendations have come to him for his work on weather, including the honorary degree of Doctor of Science, granted in recognition of his leadership in the study of hurricanes.

His hobbies continue the same as in his boyhood—watching the birds, the stars and the weather.